⋆ <u>*My 2 Cents on Calculus 1*</u> ⋆

⌊*Volume I of I-III; A Calculus Series..*)

.

⋆ ***From limits*** $\Rightarrow$ ***derivatives*** $\Rightarrow$ $u-$***sub*** ⋆

$$\left(\lim_{x\to x_0} f(x) \Rightarrow \frac{d}{dx}[f(x)] \Rightarrow \int f(x)dx\right]$$

Professor Hossien Sahebjame, M.S. Mathematics
(↓ *Long Beach, CA* ⇔ *Pasadena, TX* ↑)

2025-12-11

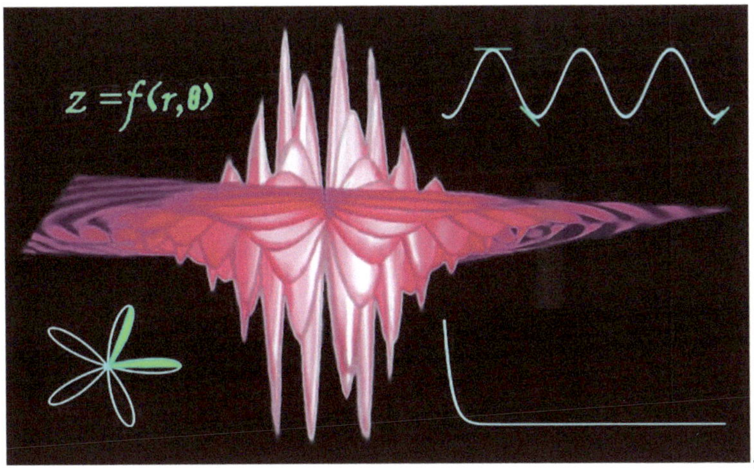

.

(Image by *Kevin Smith* (https://kevinsmithmathprog.wixsite.com/laso), *M.S. Mathematics)*

A *MyMathYourMath* (https://MyMathYourMath.com) *Project...Mathematics, for all mankind!!*

≫To the Reader/Student:

In this book I spend a very little bit of time covering things from precalc you will need to know for this textbook. I myself am a professor of mathematics and have a BS in math from UCLA and an MS in math from CSULB both in pure math. I took my masters exams in Measure & Topology. I did 2 years of a PhD math at the University of Houston but withdrew for personal reasons but finishing it. I hope this book serves you well as a first course in calculus. Typically I'm on my unicycle, in the water, or with my beautiful girlfriend Sua and her three beautiful daughters.

≫Dedicated to:

I would like to dedicate this text book to my family; my sister Maral Sahebjame, my mother Jamileh Sahebjame, my father Mohsen Sahebjame, and my girlfriend Sua Escalona and her daughters for these people are the cornerstones to my success. I could not be where I am today without them. I would lastly like to thank fellow mathematician Kevin Smith, M.S., for his many contributions and constant feedback for improvement.

≫What makes this text different than others?

I have aimed to make this textbook as self-contained as possible. Moreover, I have reformulated some key theorems for more clear understanding by the students. Also, for transcendental functions most texts first show $\frac{d}{dx}[e^x] = e^x$ then show how to differentiate say e^{2x^2-x}. I condense this into step with the formula for the generalized transcendental into a theorem. For a prelim chapter I have only focused on those topics which we use in calculus. The last section has solutions to all exercises and full solutions available in another text! Many new formulas formulated which are not conventional in typical calculus texts, they include but are not limited to: the generalized derivative for $\ln(ax^n)$ and the general formula always used for ladder sliding down a wall related rates exercise. Example answers are boxed throughout.

≫Key: Color and font codes:

1. this color means a formula not found in any other calculus textbook which I innovated.
2. this color references another section of the book.
3. this color indicates a link to a webpage for reference. In hard copies, link is provided.
4. *this* font means either a side note when its all in the font or a definition if just a word.
5. On the Contents, the numbered sections rows, 1,2,3,... you click on description to send to that section, for the lettered, (a),(b),(c),.. you must click the actual letter.

≫Contact/Community:

In my spare time, I own and operate the domain name, YouTube math Channel, Discord server by the name MyMathYourMath. I also regularly ask/answer on Quora and MathStack Exchange !

≫ Links to this colored words throughout:

1. YouTube: https://youtube.com/MyMathYourMath
2. Discord: https://discord.gg/KZ7Yp6SU
3. Quora: https://quora.com/profile/MyMathYourMath
4. MSE: https://math.stackexchange.com/users/620355/MyMathYourMath?tab=profile
5. Heine-Borel Theorem: https://en.wikipedia.org/wiki/Heine–Borel_theorem
6. Homology: https://en.wikipedia.org/wiki/Homology_(mathematics)
7. Cohomology: https://en.wikipedia.org/wiki/Cohomology
8. Metrics: https://en.wikipedia.org/wiki/Metric_space
9. Invariance of Domain: https://en.wikipedia.org/wiki/Invariance_of_domain
10. Algebraic Geometry: https://en.wikipedia.org/wiki/Algebraic_geometry
11. Separation Axioms: https://en.wikipedia.org/wiki/Separation_axiom
12. Higher sphere homotopy groups: https://en.wikipedia.org/wiki/Homotopy_groups_of_spheres
13. Absolute value function: https://en.wikipedia.org/wiki/Absolute_value
14. Algebra https://en.wikipedia.org/wiki/Algebra
15. Calculus https://en.wikipedia.org/wiki/Calculus

>§ 0. Do's & Dont's of Mathematics

Before we begin, here is a list of do's and dont's for mathematics:

1. If we write $x \in A$ it means x is an element and belongs to the set A. I.e., if $x \in (a, b)$, then $a < x < b$. So $\in$ means *is an element of* and $\notin$ means *is not an element of.* I.e., $1 \in (0, 2)$ but $1 \notin (2, 3)$.

2. Often times you think you can break up a radical or not FOIL out so
$$\sqrt{x+y} \neq \sqrt{x} + \sqrt{y}.$$
And
$$(x+y)^2 \neq x^2 + y^2.$$
But rather
$$(x+y)^2 = (x+y)(x+y) = x^2 + 2xy + y^2.$$
However, the following holds:
$$\sqrt{xy} = \sqrt{x}\sqrt{y}.$$

3. If you have $-(x+y)$ this becomes $-x-y$.

4. If A, B are two sets, their (set-theoretic) *intersection* is
$$A \cap B = \{x : x \in A \text{ and } x \in B\},$$
while their (set-theoretic) *union* is
$$A \cup B = \{x : x \in A \text{ or } x \in B\}.$$
Lastly for set theoretic operation, we have their *difference of sets* to be
$$A \setminus B = \{x \in A : x \notin B\}.$$
Often times if we have a parent or ambient space, call it X, then the difference of sets or *complement* of A in X is also denoted
$$A^c = X \setminus A.$$

5. If We write $A \subset B$ this means if $x \in A$ then $x \in B$ but if $y \in B$ then y is not necessarily in A. We say in this case A *is a subset of* B.

6. Note,
$$\frac{a}{b} + \frac{c}{d} \neq \frac{a+c}{b+d}$$
but you need to find a common denominator first then add.

7. $\sqrt[n]{x} \notin \mathbb{R}$ when n is even and $x < 0$. But if n is odd, then $\sqrt[n]{x} \in \mathbb{R}$ for any value of x.

Contents:

> § 1. Preliminaries

As I have mentioned in chapter 0, $x \in (a, b)$ implies

$$a < x < b.$$

And thus $x \in [a, b]$ implies

$$a \leq x \leq b.$$

Before moving into the pre-calc materials, I will introduce the number systems. They are:

1. $\mathbb{N} = \{1, 2, 3, ...\}$ the *natural numbers.* These are basically the counting numbers. One may argue that they are $\mathbb{N} = \{0, 1, 2, ...\}$ in which we include 0 as a natural number. We will always indicate.
2. $\mathbb{Z} = \{..., -2, -1, 0, 1, 2, ...\}$ the *integers.* And for instance $\mathbb{Z}^+ = \{1, 2, 3, ...\}$ the positive integers or $\mathbb{Z}^{\geq 0} = \{0, 1, 2, ...\}$ the non-negative integers. So integers are counting numbers, together with 0 and their negatives.
3. $\mathbb{Q} = \{\frac{p}{q} : p, q \in \mathbb{Z}\}$ the *rationals.* Basically the set of all reduced fractions. For instance, π and $\sqrt{2}$ are not in $\mathbb{Q}$.
4. $\mathbb{R} = (-\infty, \infty)$ the *reals.* The rationals, $\mathbb{Q}$ union together with the irrational numbers. I.e., the numbers with a never ending random decimal expansion like $\pi, e, \sqrt{2}$.
5. $\mathbb{C} = \{x + iy : x, y \in \mathbb{R}\}$ the *complex.* Here it is understood that $i^2 = -1$.

For the sake of our class we will not be dealing with set (5). We have the following set inclusion:

$$\mathbb{N} \subset \mathbb{Z} \subset \mathbb{Q} \subset \mathbb{R} \subset \mathbb{C}.$$

The following joke for membership in the above sets always sets the tone:

This is because $\pi \in \mathbb{R}$ but $\pi \notin \mathbb{Q}$ and $i \in \mathbb{C}$ but $i \notin \mathbb{R}$. Now, there are a few things to know from pre-calculus before beginning a first course in calculus. They are

1. Finding the equation of a line.
2. Working with exponents.

3. Evaluating functions at points.
4. Finding domains.
5. Logs and exponentials.
6. Rational functions.
7. Inequalities.

So for (1), recall, the equation of a *line* in *slope-intercept* form is given by

$$y = mx + b \qquad (*),$$

where m is the slope, and b is the $y-$intercept which is of the form $(0, b)$. The $y-$*intercept* is what you get by setting $x = 0$, and $x-$*intercept* you get by setting $y = 0$. If you're given some slope m, which for two points $(x_1, y_1), (x_2, y_2)$ is given by

$$m = \frac{\text{rise}}{\text{run}} = \frac{y_2 - y_1}{x_2 - x_1},$$

and if you have the two points, say x, y, to find b you plug in m, x, y into $(*)$ to find b and your final answer has numbers in place of m, b.

If

1. $m > 0$, line increasing.
2. $m < 0$, line decreasing.
3. $m = 0$, line horizontal.
4. m undefined, line vertical.

As below:

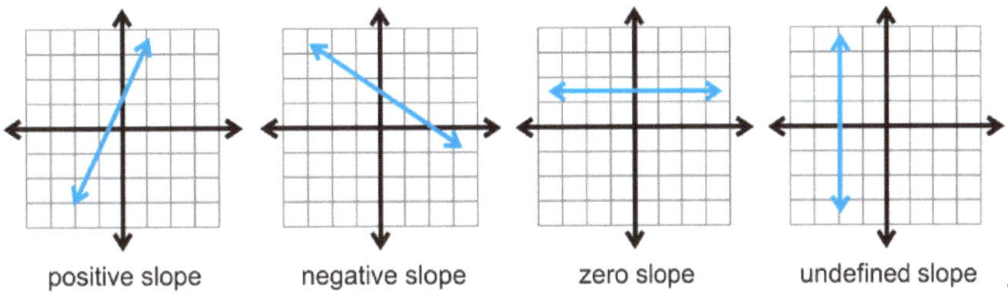

Example 1.1 Find the equation of the line passing through $(1, 4)$ with a slope of $m = 2$.

Solution 1.1 We know right off the bat that $y = mx + b$ where $m = 2, x = 1, y = 4$ so

$$\begin{aligned}
&y = mx + b; m = 2, x = 1, y = 4 \\
&\Rightarrow 4 = (2)(1) + b \\
&\Rightarrow 4 = 2 + b \\
&\Rightarrow 4 - 2 = b \\
&\Rightarrow b = 2
\end{aligned}$$

so the equation of our line is thus

$$\boxed{y = 2x + 2}.$$

A quick note on parallel and perpendicular lines: if

$$y = mx + b$$

is any line, there is an infinite number of lines both parallel to and perpendicular to y. For *parallel*, we denote by $y_{||}$ and to find it, we merely change the b intercept. I.e.,

$$y_{||} = mx + c; \qquad b \neq c$$

is parallel to y. For *perpendicular*, denoted often by $y_\perp$ we have that the $y-$intercept can be anything but the slope is now opposite and reciprocal. I.e.,

$$y_\perp = -\frac{1}{m}x + b.$$

Below are parallel, perpendicular and neither line. Here they are perpendicular:

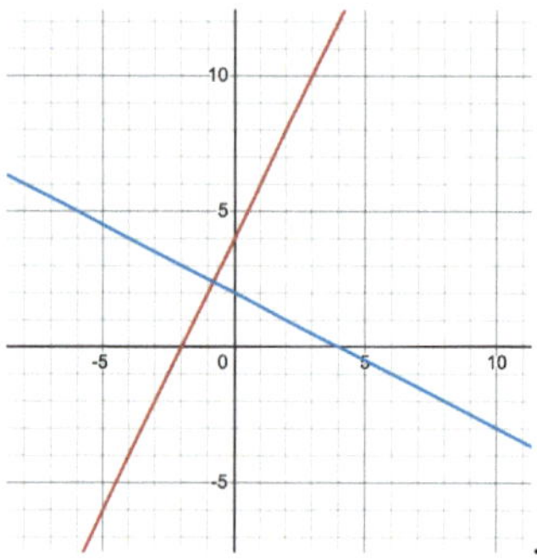

In this they are parallel:

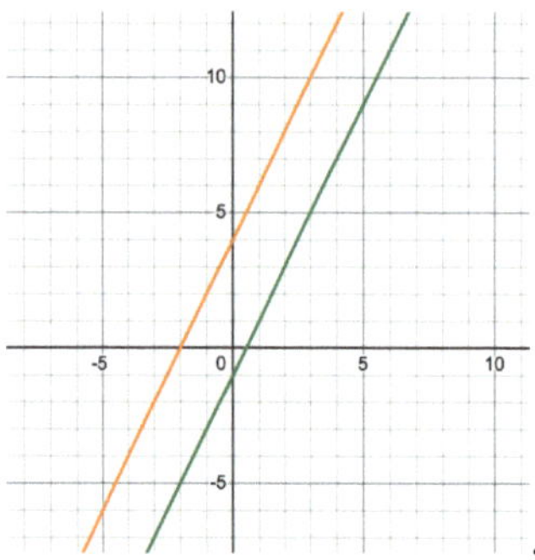

Lastly for neither perpendicular nor parallel:

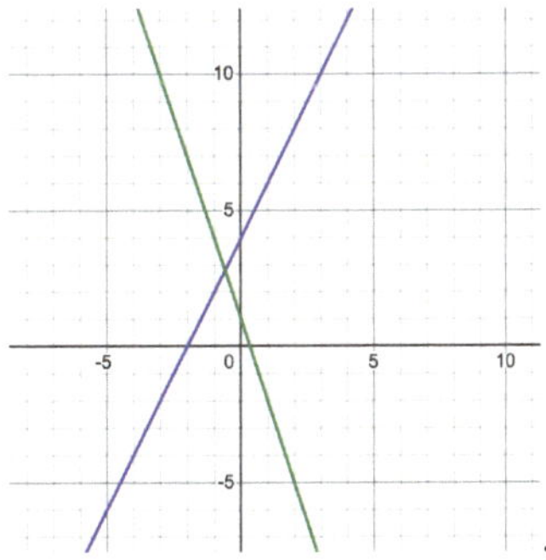

.

For (2), exponents have the following rules: (here, $a, b, c \in \mathbb{Q}$ are rationals, $x, y \in \mathbb{R}$ are real numbers.)

1. $x^0 = 1$, $x^1 = x$.
2. $x^a x^b = x^{a+b}$.
3. $(x^a)^b = x^{ab}$. More generally, $(x^a y^b)^c = x^{a \cdot c} y^{b \cdot c}$.
4. $x^a = \frac{1}{x^{-a}}$. Here of course, $x \neq 0$.
5. $x^{-a} = \frac{1}{x^a}$. Here of course, $x \neq 0$.
6. $\sqrt[n]{x^a} = x^{\frac{a}{n}} = (\sqrt[n]{x})^a$. Here for even n, we require that $x \geq 0$. If $x < 0$, the result is a complex number.

A *rational expression* is a fractional expression with one or more variables and numbers and exponents. A *radical expression* is one with variables inside the rational valued exponent, i.e., a radical.

Example 1.2(a) Simplify

$$\frac{(2x^2y^3z^4)^2}{(2x^3y^2z)^3}.$$

Solution 1.2(a) We have

$$\begin{aligned}
\frac{(2x^2y^3z^4)^2}{(2x^3y^2z)^3} &= \frac{2^2x^{2\cdot 2}y^{3\cdot 2}z^{4\cdot 2}}{2^3x^{3\cdot 3}y^{2\cdot 3}z^3} && \text{rule (3) for exponents} \\
&= \frac{\cancel{4}x^4y^6z^8}{\cancel{8}x^9y^6z^3} && \text{applying the mult} \\
&= \frac{z^8z^{-3}}{2x^9x^{-4}} && \text{(4), (5) of exponent rules} \\
&= \boxed{\frac{z^5}{2x^5}} && \text{rule (1) above.}
\end{aligned}$$

Example 1.2(b) Simplify

$$\sqrt[3]{8x^3y^5z^6}.$$

Solution 1.2(b) We have

$$\begin{aligned}\sqrt[3]{8x^3y^5z^6} &= \sqrt[3]{2^3x^3y^5z^6}\\ &= \boxed{2xyz^2\sqrt[3]{y^2}}.\end{aligned}$$

The next point, number (3), say you're given some function $f(x)$ and are asked to find say $f(x+h)$, then you swap $x+h$ wherever you see and x. I.e., given

$$f(x) = a_nx^n + a_{n-1}x^{n-1} + ... + a_1x + a_0,$$

then $f(x+h)$ is just

$$f(x+h) = a_n(x+h)^n + a_{n-1}(x+h)^{n-1} + ... + a_1(x+h) + a_0.$$

Example 1.3 Given

$$f(x) = 2x^2 + 4,$$

find $f(0), f(1), f(x+h)$.

Solution 1.3 For $f(0)$,

$$\begin{aligned}f(0) &= 2(0)^2 + 4\\ &= 0 + 4\\ &= \boxed{4}.\end{aligned}$$

Next, $f(1)$:

$$\begin{aligned}f(1) &= 2(1)^2 + 4\\ &= 2 + 4\\ &= \boxed{6}.\end{aligned}$$

So for $f(x+h)$ we have to square the term out! As follows:

$$\begin{aligned}f(x+h) &= 2(x+h)^2 + 4\\ &= 2(x+h)(x+h) + 4\\ &= 2(x^2 + xh + xh + h^2) + 4\\ &= 2(x^2 + 2xh + h^2) + 4\\ &= \boxed{2x^2 + 4xh + 2h^2 + 4}.\end{aligned}$$

Below is the graph from 1.3 with the two black dots being the evaluated points!

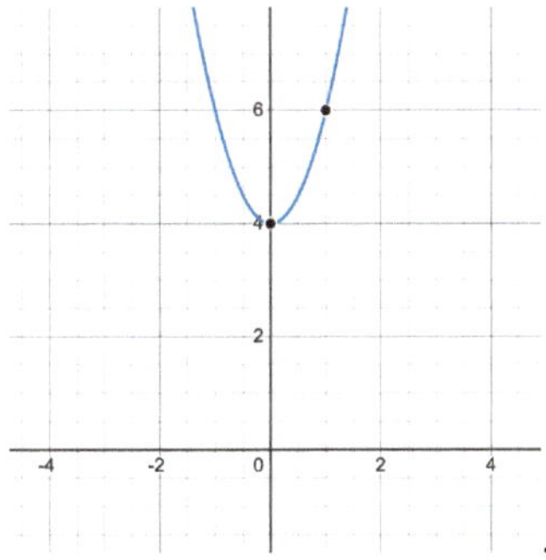

As for finding *domain*, all $x-$value inputs, except for trig functions such as $\tan x, \sec x, \csc x, \cot x$, the main cases where the domain is not $\mathbb{R}$, like in polynomials, is for example if

a) $f(x) = \frac{1}{g(x)}$ in which case domain is those x values such that $g(x) \neq 0$.

b) $f(x) = \ln(g(x))$ in which case the domain is those x values such that $g(x) > 0$.

c) $f(x) = \sqrt{g(x)}$ in which case the domain is those x values such that $g(x) \geq 0$.

For instance however, $f(x) = \tan x$ is of the form in (a) as $\tan x = \frac{\sin x}{\cos x}$ so for domain of $\tan x$ we want for

$$\cos x \neq 0 \Rightarrow x \neq \frac{\pi}{2} + \pi k; k \in \mathbb{Z}.$$

So the domain is

$$\{x \in \mathbb{R} : x \neq \frac{\pi}{2} + \pi k\}.$$

Example 1.4 Find the domain of

$$f(x) = \ln(x^2 - 1).$$

Solution 1.4 Here we are in case (b) and so we wish for

$$x^2 - 1 = (x+1)(x-1) > 0.$$

So immediately $x = \pm 1$ are not in the domain. And if $x \in (-1,1)$ say $x = 0$, we get our function is negative so the domain is

$$\boxed{(-\infty, -1) \cup (1, \infty)},$$

or

$$\boxed{\mathbb{R} \setminus [-1, 1]},$$

or

$$\boxed{\{x \in \mathbb{R} : x < -1, x > 1\}}.$$

And as you can see below, the middle portion between and including ± 1 is empty:

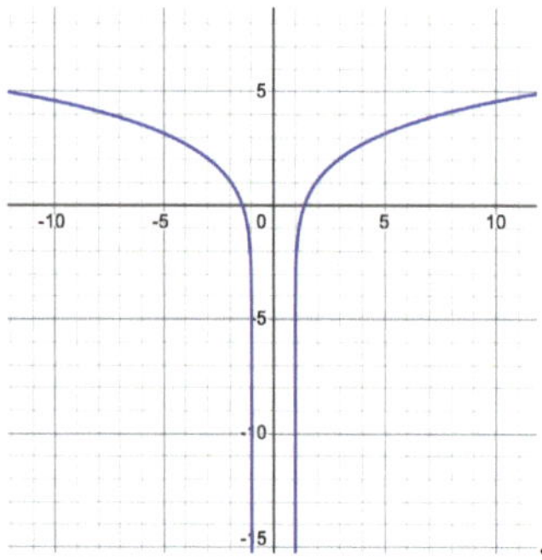

Example 1.5 Find the domain for

$$f(x) = \frac{1}{x^2 - 7x + 12}.$$

Solution 1.5 Here we want for $x^2 - 7x + 12 \neq 0$ but

$$x^2 - 7x + 12 = (x-4)(x-3)$$

so $x \neq 3, 4$. Our domain is thus

$$\boxed{(-\infty, 3) \cup (3, 4) \cup (4, \infty)},$$

or

$$\boxed{\mathbb{R} \setminus \{3, 4\}},$$

or

$$\boxed{\{x \in \mathbb{R} : x \neq 3, 4\}}.$$

Or below we note how at $x = 3, 4$ there is no graph!

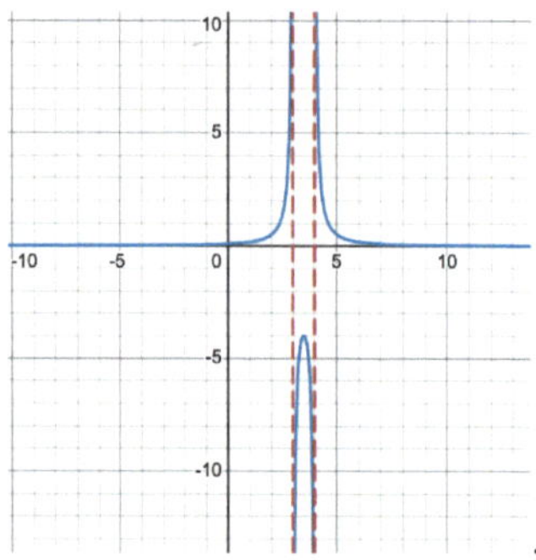

The red dotted lines denote the parts which are not in our domain here!

Example 1.6 Find the domain for

$$f(x) = \sqrt{x^2 - 4}.$$

Solution 1.6 Here, we wish for $x^2 - 4 \geq 0$ but

$$x^2 - 4 = (x+2)(x-2).$$

So $x = \pm 2$ is in our domain, and if x is between them we get negative so the domain is

$$\boxed{(-\infty, -2] \cup [2, \infty)},$$

or

$$\boxed{\mathbb{R} \setminus (-2, 2)},$$

or

$$\boxed{\{x \in \mathbb{R} : x \leq -2, x \geq 2\}}.$$

As for (5), we only really ever deal with natural log but we will cover both. First, an *exponential function* is one of the form

$$f(x) = b^x,$$

where $0 < b < 1$, or $b > 1$ and $x \in \mathbb{R}$. This has domain $\mathbb{R}$ and range $(0, \infty)$. It has inverse

$$f^{-1}(x) = \log_b x \qquad (**).$$

This is known as a *logarithmic function*. And if we let the base $b = e$ the exponential constant, then we get the *natural logarithm* or

$$\log_e x = \ln x.$$

So then it is clear the inverse for $\ln x$ is e^x. A quick note on this exponential constant, e is that one can write it as an infinite sum

$$e^x = 1 + x + \frac{x^2}{2} + \frac{x^3}{6} + \frac{x^4}{24} + \ldots$$

or in other words

$$e^x = \sum_{n=0}^{\infty} \frac{x^n}{n!}.$$

Here, $n!$ is said to be the *factorial function*, for $n \in \mathbb{Z}^{\geq 0}$, defined via

$$n! = (n)(n-1)(n-2) \cdot \ldots \cdot (n-n+1); \qquad 0! = 1.$$

The rules for logs are:

1. $\log_b b = 1$, $\log_b 1 = 0$.
2. $\log_b x + \log_b y = \log_b xy$.
3. $\log_b x - \log_b y = \log_b(\frac{x}{y})$.
4. $\log_b x^n = n \log_b x$.
5. $\log_b x = y$ means $b^y = x$.
6. From (5) we deduce, and by (**), $\log_b(b^x) = x = b^{\log_b x}$.

7. $\ln 1 = 0, \ln e = 1$.

8. $\ln(e^x) = x = e^{\ln x}$.

Example 1.7 Solve for x in

$$5\log_3(x+1) - 5\log_3(x+2) + \log_2 4 = 12.$$

Solution 1.7 We have

$$\begin{aligned}
&5\log_3(x+1) - 5\log_3(x+2) + \log_2 4 = 12 \\
&\Rightarrow 5\log_3(x+1) - 5\log_3(x+2) + 2 = 12 && \text{by (5) as } 2^y = 4 \text{ means } y = 2 \\
&\Rightarrow 5\log_3(x+1) - 5\log_3(x+2) = 10 && \text{subtract 2 from both sides} \\
&\Rightarrow \log_3(x+1) - \log_3(x+2) = 2 && \text{dividing all terms by a 5} \\
&\Rightarrow \log_3(\frac{x+1}{x+2}) = 2 && \text{by (3)} \\
&\Rightarrow \frac{x+1}{x+2} = 3^2 = 9 && \text{by (5)} \\
&\Rightarrow \frac{x+1}{x+2} - \frac{9x+18}{x+2} = 0 \\
&\Rightarrow \frac{-8x-17}{x+2} = 0 \\
&\Rightarrow x = \boxed{-\frac{17}{8}}.
\end{aligned}$$

Second to lastly, for rational functions: A *rational function*, is any function of the form

$$f(x) = \frac{p(x)}{q(x)}; \qquad q(x) \neq 0.$$

Here, $p(x), q(x)$ are polynomials. If

$$f(x) = a_n x^n + a_{n-1}x^{n-1} + \ldots + a_1 x + a_0,$$

then a_0 is the *constant term*, n is the *degree* of the polynomial, a_j for $0 \leq j \leq n$ are *coefficients*, and a_n is the *leading coefficient*. I.e., if

$$f(x) = 2x^3 - x^2 + 4x - 2,$$

the leading coefficient is 2, the constant term is -2 and the degree is 3. Rational functions have a few things to know:

- Asymptotes, vertical (we will use VA for short), horizontal (HA for short), and slant. if they exist.
- End behavior. That is which direction the function heads as your inputs head to $\pm\infty$.
- $x, y-$intercepts if any.

- If there is any symmetry.

Before we begin with the math, Asymptotes, are straight lines in our case in the graphs which the graph never crosses or touches. Really briefly, the $x-$ intercept is of the form

$$(m, 0)$$

where $x = m$ is the value obtained by setting $y = 0$ and solving for x and the $y-$ intercept is of the form

$$(0, n)$$

where $y = n$ is the value obtained by setting $x = 0$ and solving for y. For symmetries, we have two kinds, but first we must define what even and odd functions are. A function $f(x)$ is said to be *even* if

$$f(-x) = f(x)$$

holds for all x in its domain. A function $f(x)$ is said to be *odd* if

$$f(-x) = -f(x)$$

holds for all x in its domain. The two rules then are:

1. Even functions are symmetric about the $y-$axis; which means we have $(x, y) \mapsto (-x, y)$.
2. Odd functions are symmetric about the origin; which means we have $(x, y) \mapsto (-x, -y)$.

An example of an even function is x^2 because

$$\begin{aligned} f(-x) &= (-x)^2 \\ &= (-x)(-x) \\ &= x^2 \\ &= f(x). \end{aligned}$$

Thus $f(-x) = f(x)$. And if $f(x) = x^3$, it is odd because

$$\begin{aligned} f(-x) &= (-x)^3 \\ &= (-x)(-x)(-x) \\ &= x^2(-x) \\ &= -x^3 \\ &= -f(x) \end{aligned}$$

Thus $f(-x) = -f(x)$.

Side-Note: As for $\sin x, \cos x$ note $\sin x$ is odd and $\cos x$ is even. And all other 4 except for $\sec x$ are all odd.

It is worth mentioning that if f has inverse f^{-1}, then these two functions are symmetric about the line $y = x$. I.e.

$$(x, y) \mapsto (y, x)$$

is the transformation of a function to its inverse and back. Now for our key rules:

1. Vertical asymptotes are $x = a$ where $q(a) = 0$, where $q(x)$ is our denominator function.
2. Degree of numerator is less than degree of denominator. In this case: There is an HA at $y = 0$ and for end behavior:
$$f(x) \to 0; x \to \pm\infty.$$
3. Degrees are equal. Here: There is a HA at $y = \frac{a}{b}$ where a, b are leading coefficients for p, q respectively. And for end behalves,
$$f(x) \to \frac{a}{b}; x \to \pm\infty.$$
4. Degree of numerator is greater than that of denominator. Here there is no HA but rather a slant, and what we do is do long division $p(x) \div q(x)$ and the quotient of your solution is the slant. End behavior is
$$f(x) \to \infty \text{ when } x \to \infty \text{ and } f(x) \to -\infty \text{ when } x \to -\infty$$
if the fraction is negative coefficient or
$$f(x) \to -\infty \text{ when } x \to \infty \text{ and } f(x) \to \infty \text{ when } x \to -\infty$$

Example 1.8(a) Find any asymptotes and end behavior and symmetries for
$$f(x) = \frac{x^3}{x^2 - 1}.$$

Solution 1.8(a) For this degree on top is greater so there is no horizontal but rather a slant and
$$x^3 \div x^2 - 1 = x + \frac{x}{x^2 - 1}$$
and so there is a slant at $\boxed{y = x}$. And vertical asymptotes where $x^2 - 1 = 0$ which is $\boxed{x = \pm 1}$. And as for end behavior,
$$\boxed{f(x) \to \infty; x \to \infty} \text{ and } \boxed{f(x) \to -\infty; x \to \infty}$$
And for symmetries, note,
$$f(-x) = -\frac{x^3}{x^2 - 1} = -f(x)$$
so our function is odd thus symmetric about origin. Below is the graph:

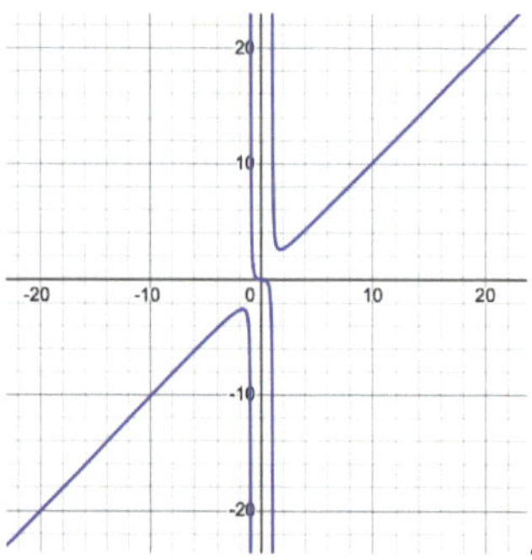

Example 1.8(b) Find any asymptotes and end behavior and symmetries for

$$f(x) = \frac{3x^2}{2x^2 + 1}.$$

Here the denominator is not zero for real values, so no VA but for HA since degrees are equal to its $\boxed{y = \frac{3}{2}}$ thus for end behavior,

$$\boxed{f(x) \to \frac{3}{2}; x \to \pm\infty}.$$

Furthermore,

$$f(-x) = \frac{3(-x)^2}{2(-x)^2 + 1} = .$$

$$\begin{aligned} f(-x) &= \frac{3(-x)^2}{2(-x)^2 + 1} \\ &= \frac{3x^2}{2x^2 + 1} \\ &= f(x) \end{aligned}$$

So it is even and thus symmetric about the $y-$axis. Below is the graph:

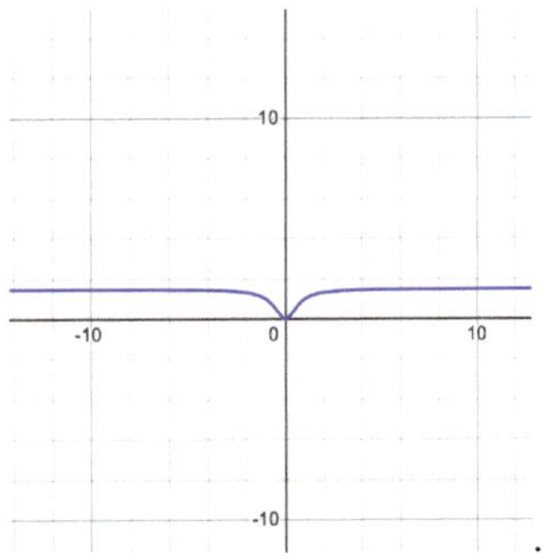

Example 1.8(c) Find any asymptotes and end behavior and symmetries for

$$f(x) = \frac{4}{x - 2}.$$

Immediately, the degree on top is lesser thus $\boxed{y = 0}$ is HA and the denominator is zero when $\boxed{x = 2}$ our VA. And thus for end behaviors,

$$\boxed{f(x) \to 0; x \to \pm\infty}.$$

Below is the graph:

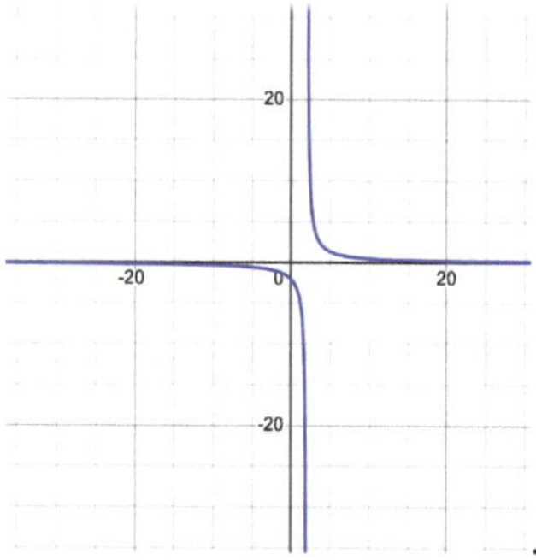

Side-Note: In the next chapters when discussing *limits*, the limit of a function has a lot to do with HA in rational functions. There is a huge connection here.

As for (7) one has (*laws of inequalities*)

1. $|x| < a$ means $-a < x < a$.
2. $|x| \leq a$ means $-a \leq x \leq a$.
3. $|x| > a$ means either $x < -a$ or $x > a$.
4. $|x| \geq a$ mean $x \geq a$ or $x \leq -a$.
5. $|x| = a$ means $x = a$ or $x = -a$.

Example 1.9 Solve for x:

$$|2x - 3| \leq 5.$$

Solution 1.9 Here by (2) on our list above,

$$\begin{aligned}
&-5 \leq 2x - 3 \leq 5 \\
&\Rightarrow -5 + 3 \leq 2x \leq 5 + 3 \\
&\Rightarrow -2 \leq 2x \leq 8 \\
&\Rightarrow -1 \leq x \leq 4 \\
&\Rightarrow \boxed{-1 \leq x \leq 4}.
\end{aligned}$$

So our answer is $\boxed{[-1, 4]}$.

>§ Outro for Chapter 1

Laws of the real numbers: For all $x, y, z \in \mathbb{R}$, one has

1. $x + y = y + x$. (*commutative property for addition*).
2. $xy = yx$. (*commutative property for multiplication*).
3. $(x + y) + z = x + (y + z)$. (*associative property of addition*).
4. $(xy)z = x(yz)$. (*associative property of multiplication*).
5. $x(y + z) = xy + xz$. (*distribution of multiplication over addition*).
6. $x + 0 = x$. (*identity property for addition*).
7. $x + (-x) = 0$. (*inverse property for addition*).
8. $x \cdot 1 = x$. (*identity property for multiplication*).
9. $x(\frac{1}{x}) = 1$. (*inverse property for multiplication*).

It should be worth noting that the *additive identity is* 0 whilst the *multiplicative identity* is 1. These laws above are also the axioms of what is known as a *vector space*, which we will not cover.

Functions: A *function*, is a mapping

$$f : X \to Y,$$

in which no two distinct elements in X do not get mapped to a single element in Y. Or more formally a *well-defined* mapping or function is one in which, if $x_1 = x_2 \in X$, then

$$f(x_1) = f(x_2) \in Y.$$

The *domain* of a function is the set of x values you are allowed to plug in. The *range*, or target space, is the set of all possible outcomes. A functions is said to be *1-1* or *injective* if for all $x_1, x_2 \in X$ with

$$x_1 \neq x_2,$$

one is guaranteed that

$$f(x_1) \neq f(x_2).$$

And it is said to be *onto* or *surjective*, if for every $y \in Y$, there exists some $x \in X$ such that

$$f(x) = y.$$

A function which is both injective and surjective is said to be a *bijection*. A function f has *inverse function*, g if for all x in the domain,

$$f(g(x)) = x = g(f(x)).$$

I.e., e^x and $\ln x$ are inverses.

A Basic review of complex numbers. A complex number, recall is any number of the form

$$z = a + bi; \quad \text{where } a, b \in \mathbb{R},\ i = \sqrt{-1}.$$

The *complex conjugate* of $z = a + bi$, is $\overline{z} = a - bi$. The *real part* of z, is $\text{Re}(z) = a$ and the *imaginary part* is $\text{Im}(z) = b$. A few facts about complex numbers:

1. If $z = a + bi, w = c + di$, then $z + w = (a + c) + (b + d)i$.
2. If $z = a + bi, w = c + di$, then $z \cdot w = (ac - bd) + (ad + bc)i$. Just FOIL.
3. If $z = a + bi$, then $z^2 = a^2 - b^2 + 2abi$.
4. $|z| = \sqrt{a^2 + b^2}$.
5. $|z \cdot w| = |z| \cdot |w|$.
6. $|z + w| \leq |z| + |w|$.
7. $|z|^2 = \overline{z} \cdot z$.
8. If $x \in \mathbb{R}$, then $|xz| = |x||z|$.

Theorem 1.1: $i^2 = -1$.

Proof. We denote $i = (0, 1)$ on the plane, thus

$$\begin{aligned} i^2 &= (0,1)^2 \\ &= (0,1)(0,1) \\ &= (0-1, 0+0) \\ &= (-1, 0). \end{aligned}$$

Thus $i^2 = -1$. □

And of course, $i^1 = i$. Then

$$i^3 = i^2 \cdot i = (-1)(i) = -i.$$

Then

$$i^4 = i^2 \cdot i^2 = (-1)(-1) = 1.$$

I promise you if you keep upping the exponent, you won't get any other solutions, save yourself the trouble and if $n = 1, 2, 3, 4, 5, 6, 7, 8.....$, then, the *powers of* i, are

$$i^n = \begin{cases} i & \text{if } n \div 4 \text{ has remainder 1} \\ -1 & \text{if } n \div 4 \text{ has remainder 2} \\ -i & \text{if } n \div 4 \text{ has remainder 3} \\ 1 & \text{if } n \div 4 \text{ has no remainder.} \end{cases}$$

Thus

$$i^{2025} = i,$$

since $2025 \div 4$ has remainder 1.

To write

$$\frac{x}{a + bi}$$

in the standard form, multiply top and bottom by the complex conjugate:

$$\begin{aligned}\frac{x}{a+bi} &= \frac{x}{a+bi}\frac{a-bi}{a-bi}\\ &= \frac{xa-xbi}{a^2+b^2}\\ &= \frac{xa}{a^2+b^2}-\frac{xb}{a^2+b^2}i.\end{aligned}$$

And similarly,

$$\begin{aligned}\frac{x}{a-bi} &= \frac{x}{a-bi}\frac{a+bi}{a+bi}\\ &= \frac{xa+xbi}{a^2+b^2}\\ &= \frac{xa}{a^2+b^2}+\frac{xb}{a^2+b^2}i.\end{aligned}$$

A basic review of trig: We must keep in mind the Pythagorean identities for trig:

1. $\sin^2 x + \cos^2 x = 1$. (Just memorize this)
2. $1 + \cot^2 x = \csc^2 x$. (To get here from (1) divide all terms by a $\sin^2 x$.)
3. $\tan^2 x + 1 = \sec^2 x$. (To get here from (1) divide all terms by a $\cos^2 x$.)

And the reciprocal and tan, cot identities:

1. $\tan x = \frac{\sin x}{\cos x}$.
2. $\cot x = \frac{\cos x}{\sin x}$.
3. $\sin x = \frac{1}{\csc x}$, $\csc x = \frac{1}{\sin x}$.
4. $\cos x = \frac{1}{\sec x}$, $\sec x = \frac{1}{\cos x}$.
5. $\tan x = \frac{1}{\cot x}$, $\cot x = \frac{1}{\tan x}$.

Side-Note: $\sin x$ and $\cos x$ are not inverses but rather *cofunctions* which means they differ by 90° That is $\sin x = \cos(90 - x)$. The inverse of say $\sin x$ is actually $\sin^{-1} x$ or $\arcsin x$ so that $\sin(\sin^{-1}(x)) = x = \sin^{-1}(\sin(x))$. So that the inverse of cos is $\cos^{-1} x$ or $\arccos x$.

Below is the *unit circle*:

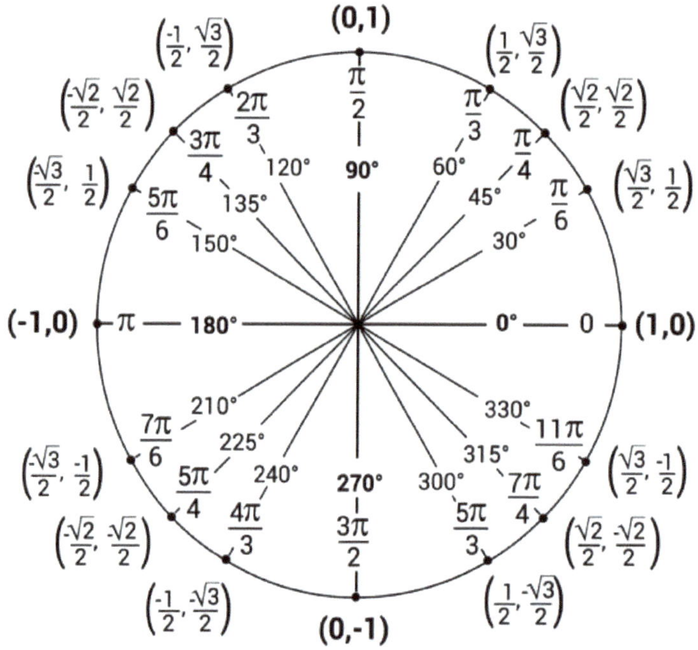

>§ 1. Exercises

1. Find the equation of a line passing through $(-2, 4)$ and $(2, 12)$.
2. Find the equation of the line with slope $m = \frac{3}{2}$ and passes through $(2, -1)$.
3. Simplify
$$\frac{(3xy^2z^3)^4}{(3x^4y^3z)^3}.$$
4. Simplify
$$\frac{(4x^4yz^2)^2}{(2xy^2z^3)^5}.$$
5. Given
$$f(x) = x^3 - x,$$
find $f(0), f(2), f(x+h)$.
6. Find the domain of
$$f(x) = \sqrt{x^2 - 9}.$$
7. Find the domain of
$$f(x) = \ln(x^2 + 3x + 2).$$
8. Find the domain of
$$f(x) = \frac{1}{2x - 3}.$$
9. Condense
$$2\log_3 x + \log_3 y - 3\log_3 z.$$
10. Expand
$$\log_4(\frac{x^2y}{z^5}).$$
11. Find the asymptotes of
$$f(x) = \frac{3}{x + 4}.$$
12. Find the asymptotes of
$$f(x) = \frac{3x^2}{2x^2 - 8}.$$
13. Find the asymptotes of
$$f(x) = \frac{x^3}{x^2 + 1}.$$
14. Solve for x:
$$|3x - 4| \geq 2.$$

>§ 2. Limits & things

>§ 2.1: Limits; Numerically & Analytically

The concept of the limit of a function is already mentioned for limits at infinity in pre-calculus when you discuss rational functions and their asymptotes. The *limit* of a function $f(x)$ as x approaches either infinity, negative infinite, or some fixed real number is a real number the function values approach and can get arbitrarily close to and is denoted

$$\lim_{x \to x_0} f(x) = L.$$

This means the limit of $f(x)$ as x approaches x_0 is the real number L.

Definition 2.1: More formally this means, if L is the *limit* of a function $f(x)$ at some x_0, then for any real numbers $\epsilon > 0$ (for any arbitrarily small positive real number), there exists some $\delta > 0$ (another arbitrarily small positive real number but this is dependent on ϵ) such that if

$$|x - x_0| < \delta,$$

then

$$|f(x) - L| < \epsilon.$$

Now limits may not always exist. For instance the limit of $f(x) = x$ as $x \to \infty$ is ∞. In this case we say, for any L, there is an $\epsilon > 0$ such that for all $\delta > 0$ one has

$$|x - x_0| < \delta$$

holds but then

$$|f(x) - L| > \epsilon.$$

There are a few ways to evaluate limits:

1. Graphically, by just looking at the graph and tracing your finger.
2. Numerically, for this method we plug in numbers from the left and right of the point being used to see what we get closer and closer to.
3. Analytically, this involves some type of trick, such as multiplication by conjugate or dividing out.

Example 2.1 Evaluate the limit

$$\lim_{x \to 0} \frac{x}{\sqrt{x+1} - 1}$$

both numerically and analytically.

Solution 2.1 Numerically, we have that we plug in for instance $-.1, -.01, -.001, ..$ and then $.1, .01, .001, ...$ So we obtain, given

$$f(x) = \frac{x}{\sqrt{x+1} - 1},$$

that

x	$-.1$	$-.01$	$-.001$	0	$.001$	$.01$	$.1$
$f(x)$	1.99499	1.9995	1.99995	X	2.00005	2.0005	2.00499

so we conclude the limit is 2. Next, analytically, to obtain the limit, one merely multiplies the top and bottom by the conjugate of the denominator!

$$\begin{aligned}\lim_{x\to 0}\frac{x}{\sqrt{x+1}-1} &= \lim_{x\to 0}\frac{x}{\sqrt{x+1}-1}\frac{\sqrt{x+1}+1}{\sqrt{x+1}+1}\\ &= \lim_{x\to 0}\frac{x\sqrt{x+1}+x}{x}\\ &= \lim_{x\to 0}\sqrt{x+1}+1\\ &= \boxed{2}.\end{aligned}$$

And it matches with the graph:

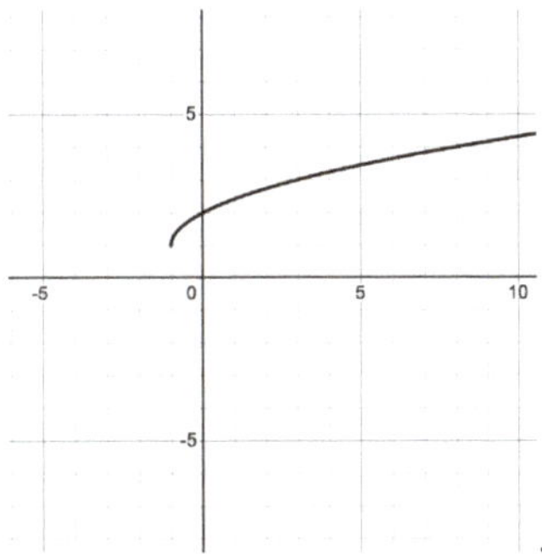

Example 2.2 The following does not exist because the left and right limits do not agree (this will become a cornerstone in limits). Show why

$$\lim_{x\to 0}\frac{|x|}{x}$$

does not exists.

Solution 2.2 Recall that

$$|x| = \begin{cases} x & x \geq 0 \\ -x & x < 0 \end{cases}.$$

This forces

$$\frac{|x|}{x} = \begin{cases} 1 & x > 0 \\ -1 & x < 0 \end{cases}.$$

Thus the values from the right approach 1 and values from left approach -1. Below you can see this graphically:

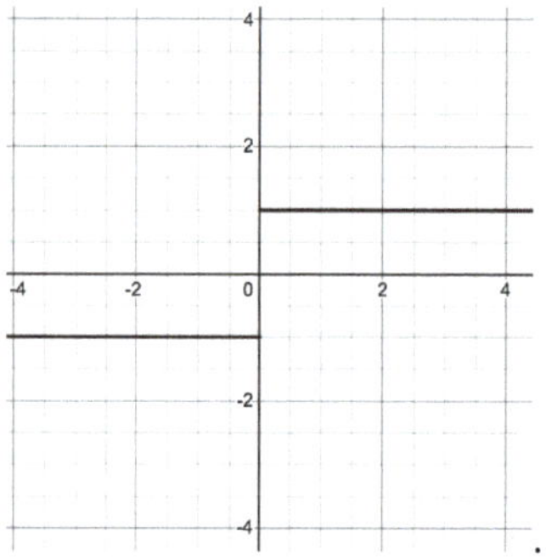

Example 2.3 Show why

$$\lim_{x \to 0} \frac{1}{x}$$

does not exist but why

$$\lim_{x \to \infty} \frac{1}{x} = 0$$

Solution 2.3 So as we approach 0 from the right we get that if x_1, x_2 are two test points with $x_1 < x_2 \in (0,1)$, then

$$\frac{1}{x_1} > \frac{1}{x_2}.$$

And from the left if $x_1, x_2 \in (-1,0)$ and $x_1 > x_2$ then

$$\frac{1}{x_1} < \frac{1}{x_2}$$

so from the right it approaches ∞ but from the left $-\infty$ thus the limit $\boxed{\text{does not exist}}$. For the limit at $x \to \infty$, we note the number gets closer and closer to zero and there is no left or right for infinities. Thus

$$\boxed{\lim_{x \to \infty} \frac{1}{x} = 0}.$$

Below is the graph so you can see

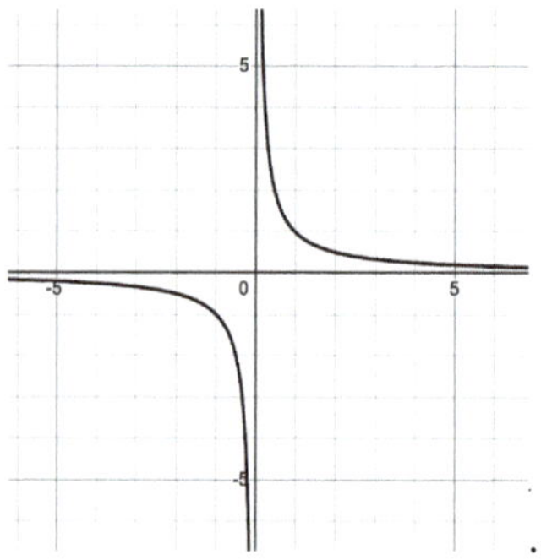

Theorem § 2.1: Basic Rules Let b, x_0 be real numbers. Let n be a positive integer. Then

1. $\lim_{x\to x_0} b = b$.
2. $\lim_{x\to x_0} x = x_0$.
3. $\lim_{x\to x_0} x^n = x_0^n$.

Theorem § 2.2: Properties of limits Let b, x_0 be real numbers. Let n be a positive integer. Suppose f, g are functions with

$$\lim_{x\to x_0} f(x) = L, \lim_{x\to x_0} g(x) = K,$$

then

1. $\lim_{x\to x_0} bf(x) = bL$.
2. $\lim_{x\to x_0}[f(x) \pm g(x)] = L \pm K$.
3. $\lim_{x\to x_0} f(x)g(x) = LK$.
4. $\lim_{x\to x_0} \frac{f(x)}{g(x)} = \frac{L}{K}$ provided $K \neq 0$.
5. $\lim_{x\to x_0}[f(x)]^n = L^n$.

Theorem § 2.3: Limits of polynomials & Rational functions Let $f(x)$ be a polynomial and x_0 a real number, then

$$\lim_{x\to x_0} f(x) = f(x_0).$$

And if $f(x) = \frac{p(x)}{q(x)}$ and $q(x_0) \neq 0$, then

$$\lim_{x\to x_0} f(x) = \frac{p(x_0)}{q(x_0)}.$$

Example 2.4 Evaluate

$$\lim_{x\to 1} 4x^2 + 3.$$

Solution 2.4 We have

$$\begin{aligned} \lim_{x\to 1} 4x^2 + 3 &= \lim_{x\to 1} 4x^2 + \lim_{x\to 1} 3 && \text{property (2)} \\ &= \lim_{x\to 1} 4x^2 + 3 && \text{Theorem 2.1} \\ &= 4(1)^2 + 3 && \text{Theorem 2.1} \\ &= \boxed{7}. \end{aligned}$$

Theorem § 2.4: Radical limits Let n be a positive integer. Then for any x_0 when n is odd and for $x_0 > 0$ when n is even we have the following:

$$\lim_{x\to x_0} \sqrt[n]{x} = \sqrt[n]{x_0}.$$

Theorem § 2.5: Limit of composite functions Let f, g be two functions with

$$\lim_{x \to x_0} g(x) = L, \lim_{x \to L} f(x) = f(L),$$

then

$$\lim_{x \to x_0} f(g(x)) = f(L).$$

Theorem § 2.6: Limits of transcendental functions Let x_0 be a real number (in the domain of the given function), then

1. $\lim_{x \to x_0} \sin x = \sin x_0$.
2. $\lim_{x \to x_0} \cos x = \cos x_0$.
3. $\lim_{x \to x_0} \tan x = \tan x_0$.
4. $\lim_{x \to x_0} \csc x = \csc x_0$.
5. $\lim_{x \to x_0} \sec x = \sec x_0$.
6. $\lim_{x \to x_0} \cot x = \cot x_0$.
7. $\lim_{x \to x_0} \ln x = \ln x_0$.
8. $\lim_{x \to x_0} a^x = a^{x_0}$.

Theorem § 2.7: Agree at all but one point functions Let x_0 be a real number and f, g two functions such that

$$f(x) = g(x)$$

except for at $x = x_0$. Then

$$\lim_{x \to x_0} f(x) = \lim_{x \to x_0} g(x).$$

Example 2.5 Evaluate the limit

$$\lim_{x \to 4} \frac{x^2 - 7x + 12}{x - 4}.$$

Solution 2.5 Note that since $f(x) = x^2 - 7x + 12 = (x - 4)(x - 3)$ and our $g(x) = x - 4$ and at $x = 4$ so by theorem 2.7:

$$\begin{aligned}
\lim_{x \to 4} \frac{x^2 - 7x + 12}{x - 4} &= \lim_{x \to 4} \frac{\cancel{(x - 4)}(x - 3)}{\cancel{(x - 4)}} \\
&= \lim_{x \to 4} (x - 3) \\
&= 4 - 3 \\
&= \boxed{1}.
\end{aligned}$$

Example 2.6 Evaluate

$$\lim_{x \to 0} \frac{\sqrt{x + 1} - 1}{x}.$$

Solution 2.6 If we just plug in $x = 0$ we get $\frac{0}{0}$. But if we multiply the top and bottom by the conjugate of the top, that is, multiply them both by

$$\sqrt{x+1}+1$$

we get

$$\begin{aligned}\lim_{x\to 0}\frac{\sqrt{x+1}-1}{x} &= \lim_{x\to 0}\frac{\sqrt{x+1}-1}{x}\cdot\frac{\sqrt{x+1}+1}{\sqrt{x+1}+1}\\ &= \lim_{x\to 0}\frac{x}{x(\sqrt{x+1}+1)}\\ &= \lim_{x\to 0}\frac{1}{\sqrt{x+1}+1}\\ &= \frac{1}{\sqrt{1}+1}\\ &= \boxed{\frac{1}{2}}.\end{aligned}$$

Below is it graphically:

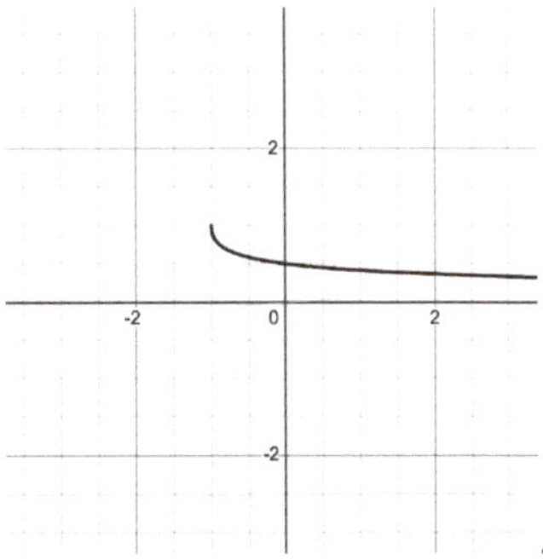

Theorem § 2.8: The squeeze theorem Suppose f, g, h are three functions such that on some interval (a, b) except for some x_0 with $a < x_0 < b$, one has

$$f(x) \leq g(x) \leq h(x).$$

Suppose in addition that

$$\lim_{x\to x_0} f(x) = L = \lim_{x\to x_0} h(x).$$

Then one has that

$$\lim_{x\to x_0} f(x)$$

exists and equals L.

Theorem § 2.9: Some special limits The following will be used usually with Squeeze theorem:

1. $\lim_{x\to 0}\frac{\sin x}{x-1}$.
2. $\lim_{x\to 0}\frac{1-\cos x}{x} = 0$.

3. $\lim_{x\to\infty}(1+x)^{\frac{1}{x}} = e$.

Example 2.7 Evaluate

$$\lim_{x\to 0}\frac{\tan x}{x}.$$

Solution 2.7 We know $\tan x = \frac{\sin x}{\cos x}$ so

$$\begin{aligned}\lim_{x\to 0}\frac{\tan x}{x} &= \lim_{x\to 0}\frac{\sin x}{x}\frac{1}{\cos x}\\ &= \lim_{x\to 0}\frac{\sin x}{x}\cdot\lim_{x\to 0}\frac{1}{\cos x}\\ &= (1)(1)\\ &= \boxed{1}.\end{aligned}$$

This is because for $\cos x$ at $x = 0$ we get 1 and we know from our previous theorem that $\frac{\sin x}{x}$ has limit of 1 as $x \to 0$. Below is the graph!

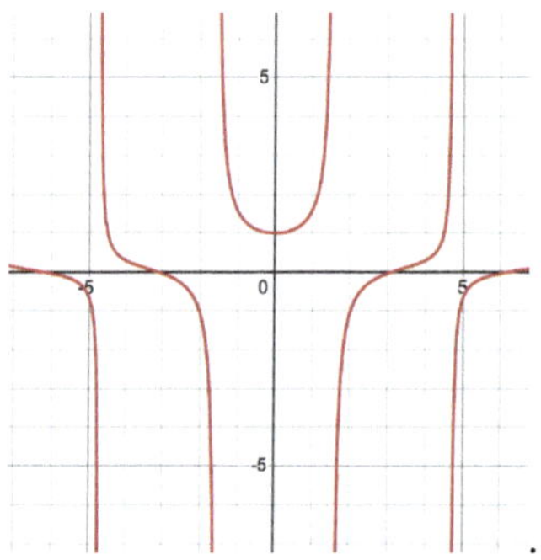

.

>§ 2.1 Exercises

1. Evaluate
$$\lim_{x\to 1} x^2 - 3x + 1.$$
2. Evaluate
$$\lim_{x\to \frac{\pi}{2}} 2\sin x.$$
3. Evaluate
$$\lim_{x\to e} \frac{\ln x}{2}.$$
4. Evaluate
$$\lim_{x\to 0} e^{x+1}.$$
5. Evaluate
$$\lim_{x\to 1} \frac{x^2 - 3x + 2}{x - 1}.$$
6. Evaluate
$$\lim_{x\to 0} \frac{\sqrt{x+25} - 5}{x}.$$
7. Evaluate
$$\lim_{x\to \infty} \frac{e^{\sin x}}{\ln(\ln(x))}.$$
8. Evaluate
$$\lim_{x\to 0} \frac{1}{\sqrt{x+1} - 1}$$
9. Show using the ϵ, δ definition of a limit that
$$\lim_{x\to 0} e^x = 1.$$
   I.e., find a δ which depends on ϵ so that the limit is satisfied.

>§ 2.2: Continuous functions & One-Sided Limits

A function being said to be *continuous* is similar to in real life. It means there is no breakage in the graph of the function. There are three conditions for a function being continuous at some $x = x_0$ point of the domain. Below is a function which is not defined at x_0.

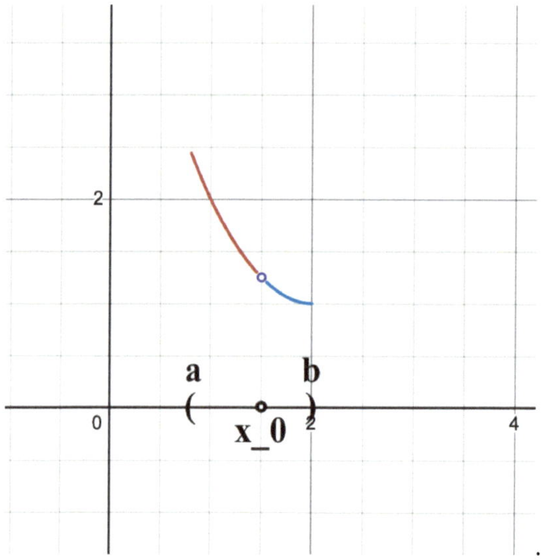

Below is a function whose limit does not exists at $x = x_0$:

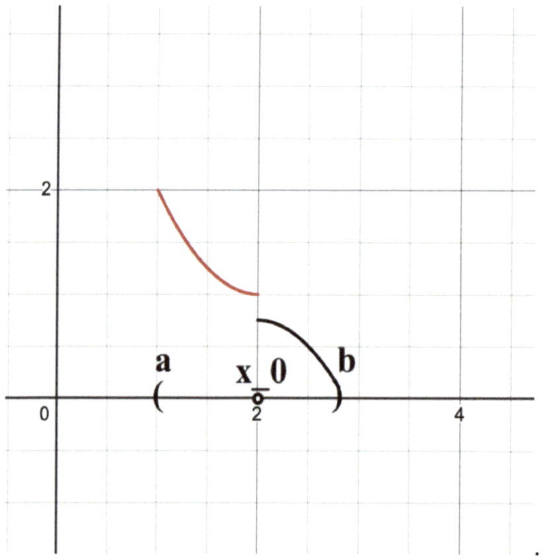

Below is when the limit of the function doesn't equal the function at the point:

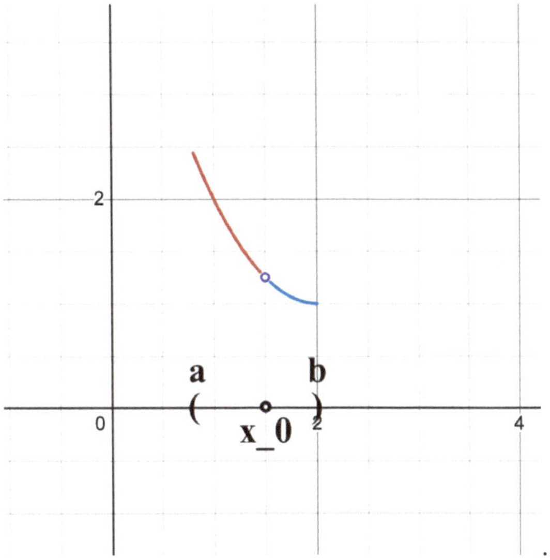

Definition 2.2: Continuous function A function $f(x)$ is said to be *continuous at some* $x = x_0$ if the following hold:

1. $f(x_0)$ is defined.
2. $\lim_{x \to x_0} f(x)$ exists and equals $f(x_0)$.

A more formal definition is a function $f : X \to Y$ is *continuous at some* $x = x_0 \in X$ if for any given $\epsilon > 0$ there exists a $\delta > 0$ such that if

$$|x - x_0| < \delta,$$

then

$$|f(x) - f(x_0)| < \epsilon.$$

If a function is continuous everywhere is just said to be *continuous.* Or if it is continuous for every $x \in (a, b)$ we say f is continuous on (a, b). otherwise we say it is *discontinuous.*

Example 2.8(a) Describe the continuity of

$$f(x) = \frac{1}{x-1}.$$

Solution 2.8(a) Since here our denominator cannot equal zero and this happens precisely when $x = 1$, thus $x = 1$ is not in our domain thus $f(x)$ is continuous at every point on the real number line except for at $x = 1$. I.e., f is continuous on

$$\boxed{\mathbb{R} \setminus \{1\}}.$$

As we can see below:

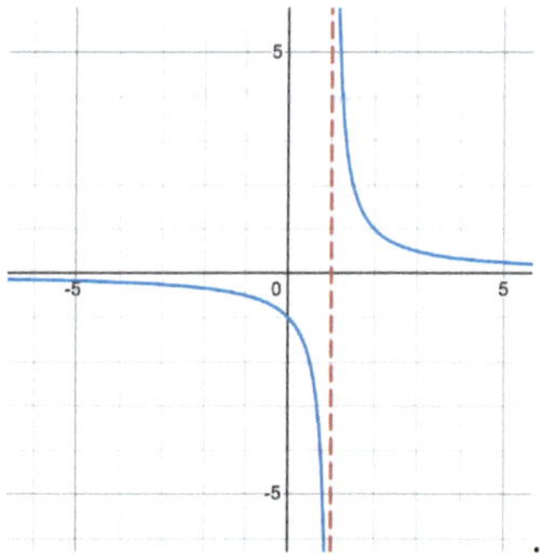

Fact 2.1: Functions that are continuous everywhere The following are continuous on the whole real line:

1. All polynomials.
2. For $f(x) = \sqrt[n]{x}$ when n is an odd positive integer. If n is even its only continuous for non-negative reals. I.e., on $[0, \infty)$.
3. $\sin x, \cos x$.
4. $f(x) = e^x$.
5. The absolute value functions.

Example 2.8(b) Describe the continuity of

$$f(x) = \tan x.$$

Solution 2.8(a) Here we know

$$\tan x = \frac{\sin x}{\cos x}$$

and if $\cos x = 0$, then $x = \frac{\pi}{2} + \pi k$ where $k \in \mathbb{Z}$. So $f(x)$ is continuous on

$$\boxed{\mathbb{R} \setminus \{\frac{\pi}{2} + \pi k : k \in \mathbb{Z}\}}.$$

or

$$\boxed{\{x \in \mathbb{R} : x \neq \frac{\pi}{2} + \pi k, k \in \mathbb{Z}\}}.$$

Or merely

$$\boxed{... \cup \left(-\frac{3\pi}{2}, \frac{\pi}{2}\right) \cup \left(\frac{\pi}{2}, \frac{3\pi}{2}\right) \cup ...}$$

Next we discuss the concept of a one-sided limits. They are denotes

$$\lim_{x \to x_0^+} f(x) = L$$

which means the limit as x approaches x_0 with values from the right of x_0 is L and if

$$\lim_{x \to x_0^-} f(x) = L,$$

then the limit as x approaches x_0 with values from the left of x_0 is L. I.e., if n is an even integer, then

$$\lim_{x \to 0^+} \sqrt[n]{x} = 0.$$

Theorem § 2.10: Existence of a Limit Let f be a function, $x_0, L \in \mathbb{R}$. Then we say $\lim_{x \to x_0} f(x) = L$ if and only if

$$\lim_{x \to x_0^-} f(x), \ \lim_{x \to x_0^+} f(x)$$

both exists and are equal to L.

Definition 2.3: Continuity on a closed interval We say a function f is continuous on $[a, b]$ if it is continuous on (a, b) and

$$\lim_{x \to a^-} f(x) = f(a)$$

and

$$\lim_{x \to b^+} f(x) = f(b).$$

Theorem § 2.11: Properties of continuous functions Let f, g be continuous at some $x = x_0$ and let $b \in \mathbb{R}$. Then the following are also continuous at $x = x_0$:

1. $bf(x)$, scalar products.
2. $f(x) \pm g(x)$, sums and differences.
3. $f(x)g(x)$, products.
4. $\frac{f(x)}{g(x)}$, $g(x) \neq 0$, quotients.

Theorem § 2.12: Continuity of composite functions Suppose g is continuous at x_0 and f is continuous at $g(x_0)$, then

$$(f \circ g)(x) = f(g(x))$$

is continuous at $x = x_0$.

Example 2.9(a) Determine the continuity of

$$f(x) = \sin\left(\frac{1}{x}\right).$$

Solution 2.9(a) We note that $\frac{1}{x}$ is continuous everywhere except at $x = 0$ so $f(x)$ is continuous everywhere except $x = 0$. I.e., is continuous on

$$\boxed{(-\infty, 0) \cup (0, \infty)}$$

or

$$\boxed{\mathbb{R} \setminus \{0\}}$$

or

$$\boxed{\{x \in \mathbb{R} : x \neq 0\}}.$$

Example 2.9(b) Determine the continuity of

$$f(x) = \begin{cases} x\sin(\frac{1}{x}) & x \neq 0 \\ 0 & x = 0 \end{cases}.$$

Solution 2.9(b) Note that since $-1 \leq \sin(\frac{1}{x}) \leq 1$, one has

$$-|x| \leq x\sin\left(\frac{1}{x}\right) \leq |x|.$$

So by the squeeze theorem, since

$$\lim_{x\to 0} -|x| = 0 = \lim_{x\to 0} |x|,$$

one has

$$\lim_{x\to 0} x\sin\left(\frac{1}{x}\right) = 0$$

thus f is continuous everywhere.

Example 2.10(a) Evaluate

$$\lim_{x\to 0^+} \frac{1}{2+2^{-\frac{1}{x}}}.$$

Solution 2.10(a) We first note that

$$\lim_{x\to 0^+} 2^{-\frac{1}{x}} = 0,$$

while

$$\lim_{x\to 0^-} 2^{-\frac{1}{x}} = \infty.$$

Thus

$$\begin{aligned} \lim_{x\to 0^+} \frac{1}{2+2^{-\frac{1}{x}}} &= \frac{1}{2+0} \\ &= \boxed{\frac{1}{2}}. \end{aligned}$$

Example 2.10(b) Evaluate

$$\lim_{x\to 0^-} \frac{1}{2+2^{-\frac{1}{x}}}.$$

Solution 2.10(b) We have that since $\lim_{x\to 0^-} 2^{-\frac{1}{x}} = \infty$ thus

$$\lim_{x\to 0^-} \frac{1}{2+2^{-\frac{1}{x}}} = \boxed{0}.$$

We can see this in the graph below:

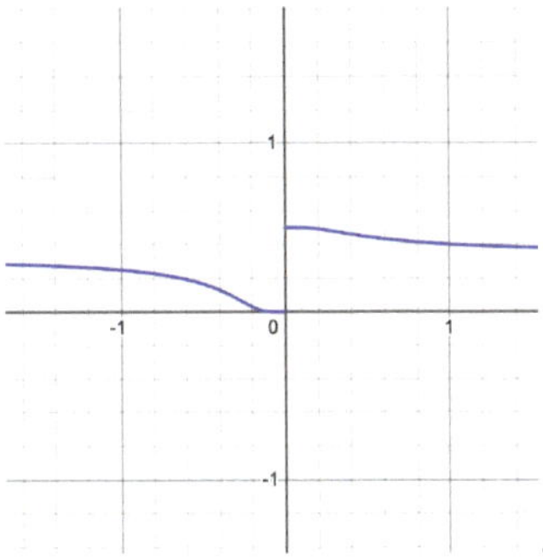

So $\lim_{x\to 0} \frac{1}{2+2^{-\frac{1}{x}}}$ $\boxed{\text{does not exist}}$ since the left and right limits do not agree.

Example 2.10(c) Evaluate

$$\lim_{x\to 2^+} \frac{|x-2|}{x-2}.$$

And

$$\lim_{x\to 2^-} \frac{|x-2|}{x-2}.$$

Solution 2.10(c) Note

$$|x-2| = \begin{cases} -(x+2) & x < 2 \\ x+2 & x \geq 2 \end{cases}.$$

Thus from the left, this approaches -1 and as $x \to 2^+$ we approach 1. As seen in the graph:

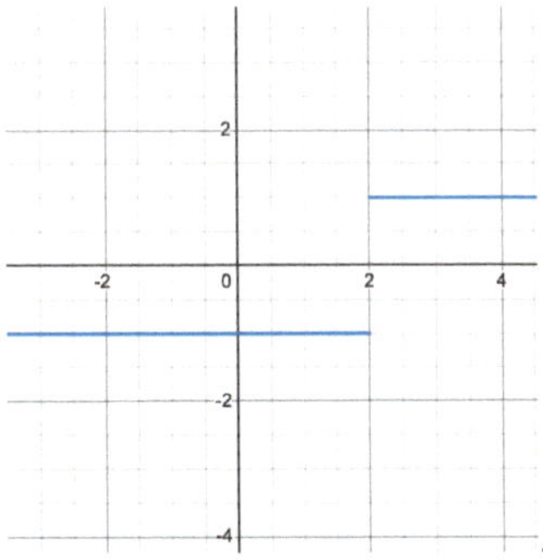

Since the left limit is $\boxed{-1}$ and the right is $\boxed{1}$, the limit $\lim_{x\to 2} \frac{|x-2|}{x-2}$ $\boxed{\text{does not exist}}$.

>§ 2.2 Exercises

1. Determine the continuity for
$$f(x) = 2x^3 - x + 1.$$

2. Determine the continuity for
$$f(x) = \frac{1}{2x-1}.$$

3. Determine the continuity for
$$f(x) = \ln(x^2 - 1).$$

4. Determine the continuity for
$$f(x) = 2\sin(2x).$$

5. Determine the continuity for
$$f(x) = e^{-x}.$$

6. Evaluate
$$\lim_{x\to 0^+} \frac{1}{4 + 2^{-\frac{1}{x}}},$$
and
$$\lim_{x\to 0^-} \frac{1}{4 + 2^{-\frac{1}{x}}}.$$

7. Evaluate
$$\lim_{x\to 0^+} \ln x,$$
and
$$\lim_{x\to 0^-} \ln x.$$

>§ **2.3: Infinite Limits** An *infinite limit* is when

$$\lim_{x\to x_0} f(x) = \pm\infty.$$

Definition 2.4: A function has an *positive infinite limit as it approaches* x_0 if for any $\epsilon > 0$ there is a $\delta > 0$ such that if

$$|x - x_0| < \delta,$$

then

$$|f(x)| > \epsilon.$$

Similarly *negative infinite limit* the $\epsilon > 0$ is replaced with $\epsilon < 0$ and it becomes

$$|f(x)| < \epsilon.$$

Theorem § 2.13: Properties of infinite limits Let $x_0, L \in \mathbb{R}$ and f, g functions with

$$\lim_{x\to x_0} f(x) = \infty, \ \lim_{x\to x_0} g(x) = L.$$

Then

1. $\lim_{x\to x_0}[f(x) \pm g(x)] = \infty$.
2. $\lim_{x\to x_0}[f(x)g(x)] = \infty$ for $L > 0$ and $-\infty$ for $L < 0$.
3. $\lim_{x\to x_0} \frac{g(x)}{f(x)} = 0$.

Example 2.11 Evaluate the limit:

$$\lim_{x\to 0}(1 + \frac{1}{x^2}).$$

Solution 2.11 We have

$$\begin{aligned}\lim_{x\to 0}(1 + \frac{1}{x^2}) &= \lim_{x\to 0} 1 + \lim_{x\to 0} \frac{1}{x^2} \\ &= 1 + \infty \\ &= \boxed{\infty}.\end{aligned}$$

Example 2.12(a) Evaluate

$$\lim_{x\to 2^-} \frac{3}{x-2}.$$

Solution 2.12(q) Note that as you plug in $1.9, 1.99, 1.999$ we get closer and closer to $-\infty$ o here it's $\boxed{-\infty}$.

Example 2.12(b) Evaluate

$$\lim_{x\to 2^+} \frac{3}{x-2}.$$

Solution 2.12(b) Similar to (a), here we plug in $2.1, 2.01, 2.001$ and we get closer to $\boxed{\infty}$.

Example 2.13 Determine if

$$\lim_{x\to 0} \frac{1}{x}$$

exists or not.

Solution 2.13 To see if it exists, we check if the left and right limits agree but by similar reasoning to Example 2.11, we have

$$\lim_{x\to 0^-}\frac{1}{x} = -\infty \neq \infty = \lim_{x\to 0^+}\frac{1}{x}$$

thus

$$\lim_{x\to 0}\frac{1}{x}$$

$\boxed{\text{does not exists}}$.

Example 2.14 Determine the limit

$$\lim_{x\to 3}\frac{1}{(x-3)^4}.$$

Solution 2.14 We must test to see if the left and right limits agree. But

$$\lim_{x\to 3^-}\frac{1}{(x-3)^4} = \infty,$$

and

$$\lim_{x\to 3^+}\frac{1}{(x-3)^4} = \infty$$

thus by theorem § 2.10, we can conclude that

$$\boxed{\lim_{x\to 3}\frac{1}{(x-3)^4} = \infty}.$$

Example 2.15 Find

$$\lim_{x\to 0} x(3+x)^2.$$

Solution 2.15 This quickly becomes

$$\lim_{x\to 0} x(3+x)^2 = \lim_{x\to 0}\frac{(3+x)^2}{\frac{1}{x}}.$$

At this point, from theorem § 2.13(3) our $f(x) = \frac{1}{x}$ which has limit ∞ as $x \to 0$ and our $g(x) = (3+x)^2 = 9$ as $x \to 0$ thus we get 0. I.e.,

$$\boxed{\lim_{x\to 0} x(3+x)^2 = 0}.$$

Though this one could have been done by direct substitution!

>§ 2.3 Exercises

1. Determine the limit, if it exists:
$$\lim_{x\to\frac{\pi}{2}} \tan x.$$

2. Determine the limit, if it exists:
$$\lim_{x\to 0} \frac{5}{x^4}.$$

3. Determine the limit, if it exists:
$$\lim_{x\to 1} \frac{x^2}{x-1}.$$

4. Determine the limit, if it exists:
$$\lim_{x\to \pi} \frac{\sin x}{x-\pi}.$$

5. Describe from a pre-calculus standpoint why
$$\lim_{x\to\infty} \frac{x^3}{x^2+1} = \infty$$
and why
$$\lim_{x\to\infty} \frac{3x^3}{2x^3-x+1} = \frac{3}{2}$$
and
$$\lim_{x\to\infty} \frac{3}{x-2} = 0.$$

>§ **2.4: Epsilon Delta proofs** In this section we learn how to, given a limit L or an $\epsilon > 0$ bound on the difference of the outputs, we learn how to formally construct our choice of delta. Basically the choice of your δ is dependent on the given arbitrary ϵ.

Example 2.15 Find the δ such that

$$\lim_{x \to 3} 2x - 5 = 1.$$

Solution 2.15 So here, we wish to find a $\delta > 0$ such that if

$$|x - 3| < \delta,$$

then

$$|f(x) - f(3)| = |2x - 5 - 1| < \epsilon.$$

For any given arbitrary $\epsilon > 0$. Typically here we work backwards: we want for

$$|2x - 6| < \epsilon.$$

But this means

$$|x - 3| > \epsilon.$$

I.e.,

$$|x - 3| < \frac{\epsilon}{2}.$$

But this is what we would like to assume except for a delta no epsilon so our choice of delta must be

$$\boxed{\delta = \frac{\epsilon}{2}}.$$

Example 2.16(a) Find the δ such that

$$\lim_{x \to 2} x^2 = 4.$$

Solution 2.16(a) We must, for any given $\epsilon > 0$, find a δ (dependent on ϵ) so that if

$$|x - 2| < \delta,$$

then

$$|x^2 - 4| < \epsilon.$$

But we know

$$|x^2 - 4| = |x + 2||x - 2|.$$

This is a property of the absolute value. We are interested in values close to 2 on $(1, 3)$ note

$$|x + 2| < 5.$$

so then we must choose $\boxed{\delta = \min\left\{\frac{\epsilon}{5}, 1\right\}}$.

Example 2.16(b) We could also go about this more directly which is my preferred method. I.e., If we wish to find δ for

$$\lim_{x\to 0}\sqrt{x}=0,$$

Solution 2.16(b) We would start with the "then" part of our conditional statement. We want a $\delta>0$ such that for any $\epsilon>0$ one has that if

$$|x|<\delta,$$

then

$$\left|\sqrt{x}\right|<\epsilon.$$

But this last statement really means

$$-\epsilon<\sqrt{x}<\epsilon.$$

But this means

$$x<\epsilon^2.$$

So we can choose $\boxed{\delta=\epsilon^2}$.

>§ 2.4: Exercises

1. Find the appropriate delta to prove
$$\lim_{x\to 3}(4-x)=1.$$
2. Find the appropriate delta to prove
$$\lim_{x\to 1}\sqrt{x}=1.$$
3. Find the appropriate delta to prove
$$\lim_{x\to 6}\frac{2}{x-5}=2.$$
4. Find the appropriate delta to prove
$$\lim_{x\to 2}\frac{1}{x-1}=1.$$
5. Find the appropriate delta to prove
$$\lim_{x\to 0}(2x+3)=3.$$
6. **Bonus** Find the appropriate delta to prove
$$f(x)=e^{x}$$
is continuous.

>§ 3. Differentiation

>§ **3.1 Differentiation & the tangent line problem.** In this section we learn how to find the equations of a line tangent to a curve at a given point, use the definition of derivative to find the derivative, and describe relationship between continuity & differentiability

Definition 3.1 If f is a given function and $x = x_0$ is in the domain, then if the following limit:

$$m = \lim_{h \to 0} \frac{f(x_0 + h) - f(x_0)}{h},$$

exists, then the equation of the line passing through $(x_0, f(x_0))$ with slope m is the *equation of the tangent line to* $f(x)$ *at* x_0. Here, n is also known as the *slope of the tangent line to the function at the given point.* This slope is often denoted $f'(x_0)$. The derivative typically indicates a certain rate of change. Typically, the size of a certain volume or length with respect to a time variable, t.

Below in blue is the tangent line and in red is the given curve:

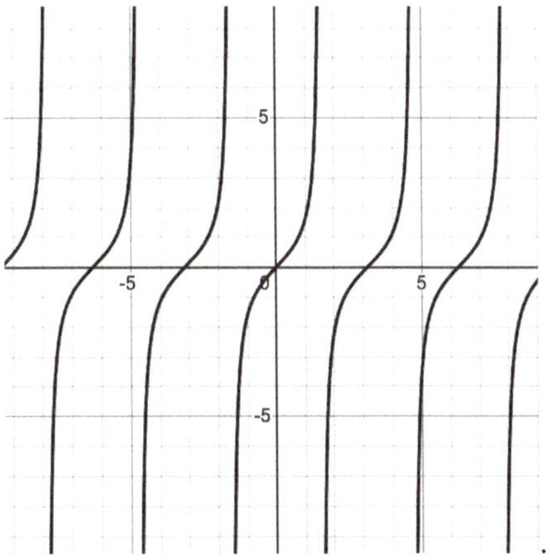

Example 3.1 Find the slope of the tangent line to

$$f(x) = 2x^2 - 3$$

at $x = 1$.

Solution 3.1 First we must find the limit above for this function:

$$\begin{aligned}\lim_{h\to 0}\frac{f(x+h)-f(x)}{h} &= \lim_{h\to 0}\frac{2(x+h)^2-3-2x^2+3}{h}\\ &= \lim_{h\to 0}\frac{2(x^2+2xh+h^2)-3-2x^2+3}{h}\\ &= \lim_{h\to 0}\frac{\cancel{2x^2}+4xh+2h^2-\cancel{3}-\cancel{2x^2}+\cancel{3}}{h}\\ &= \lim_{h\to 0}\frac{\cancel{h}(4x+2h)}{\cancel{h}}\\ &= \lim_{x\to 0}4x+2h\\ &= \boxed{4x}.\end{aligned}$$

Thus the slope at $x=1$ is $f'(1)=4(1)=\boxed{4}$.

Definition 3.2 The *derivative of a function $f(x)$ at some point x_0* is given by

$$f'(x_0)=\lim_{h\to 0}\frac{f(x_0+h)-f(x_0)}{h} \qquad (*).$$

This is read "$f'(x_0)$ *is f prime of* x_0". This is often used to calculate instantaneous rates of change. Some common notations include but are not limited to (here $f(x), y$ are synonymous)

$$f'(x_0), y', \left.\frac{dy}{dx}\right|_{x_0}, \frac{d}{dx}[f(x_0)], D_{x_0}(y).$$

If this limit exists, we say that the function $f(x)$ is differentiable at $x=x_0$. The limit in $(*)$ is also known as the *difference quotient.*

Definition 3.3 A function f is said to be *differentiable at* x_0 if the difference quotient limit exists. If it exists everywhere we say it is *differentiable.*

A list of functions and their differentiability:

1. $f(x)=x$ and all polynomials are differentiable on $\mathbb{R}$.
2. $f(x)=|x|$ is not differentiable at the origin.
3. $f(x)=\sin x$ or $\cos x$ are differentiable everywhere.
4. $f(x)=\tan x$ is differentiable everywhere except where $\cos x=0$ which is $x=\frac{\pi}{2}+\pi k,\ k\in\mathbb{Z}$.
5. $f(x)=\frac{1}{x}$ is also differentiable except for at $x=0$.

Example 3.2 Find the derivative of

$$f(x)=\frac{2}{x}.$$

Solution 3.2 We apply the difference quotient and we get

$$\begin{aligned}
\lim_{h\to 0}\frac{f(x+h)-f(x)}{h} &= \lim_{h\to 0}\frac{\frac{2}{x+h}-\frac{2}{x}}{h}\\
&= \lim_{h\to 0}\frac{\frac{2(x)}{(x)(x+h)}-\frac{2(x+h)}{x(x+h)}}{h}\\
&= \lim_{h\to 0}\frac{\frac{\cancel{2x}-\cancel{2x}-2h}{(x)(x+h)}}{h}\\
&= \lim_{h\to 0}-\frac{2\cancel{h}}{x^2+xh}\frac{1}{\cancel{h}}\\
&= \lim_{h\to 0}-\frac{2}{x^2+xh}\\
&= \boxed{-\frac{2}{x^2}}.
\end{aligned}$$

Example 3.3 Find the derivative of

$$f(x)=\sqrt{x}.$$

Solution 3.3 We have, by the difference quotient,

$$\begin{aligned}
\lim_{h\to 0}\frac{\sqrt{x+h}-\sqrt{x}}{h} &= \lim_{h\to 0}\frac{\sqrt{x+h}-\sqrt{x}}{h}\cdot\frac{\sqrt{x+h}+\sqrt{x}}{\sqrt{x+h}+\sqrt{x}}\\
&= \lim_{h\to 0}\frac{\cancel{x}+\cancel{h}-\cancel{x}}{\cancel{h}(\sqrt{x+h}+\sqrt{x})}\\
&= \lim_{h\to 0}\frac{1}{\sqrt{x+h}+\sqrt{x}}\\
&= \boxed{\frac{1}{2\sqrt{x}}}.
\end{aligned}$$

When given an equation, and an x value, to find the *equation of the line tangent to the curve at the given point*, you

1. Find the y value so if your given point is x_0, the $y-$value is $f(x_0)$.
2. Find the slope, i.e., $f'(x_0)$.
3. Find the $y-$intercept.
4. Put it all together as equation of line in slope intercept form.

We will not introduce this topic until the next section when differentiation has quick tricks.

Theorem § **3.0** If a function is differentiable on an interval, it is continuous on this interval. A proof can be found in Appendix I, pg 123..

>§ 3.1: Exercises

1. Find $f'(x)$ if
$$f(x) = 2x^2 - 4x + 5.$$

2. Find $f'(x)$ if
$$f(x) = \frac{1}{2}x^2 + x - 1.$$

3. Find $f'(x)$ if
$$f(x) = \ln x.$$

4. Find $f'(x)$ if
$$f(x) = \sin x.$$

5. Find $f'(x)$ if
$$f(x) = \sqrt{2x+1}.$$

6. Find $f'(x)$ if
$$f(x) = \frac{1}{x}.$$

>§ 3.2: Rules of differentiation

In this section, we learn about the rules for derivatives which make life simpler. But you got to crawl before you walk so you must master the definition method before learning these rules.

First and foremost, constant, they have derivative zero!

Theorem 3.1: Derivative of constant Constants have derivative zero. That is if k is constant, then if

$$f(x) = k,$$

then

$$f'(x) = 0.$$

or equivalently, $\frac{d}{dx}[k] = 0$.

Theorem § 3.2: The power rule If you're given

$$f(x) = ax^n,$$

then

$$f'(x) = anx^{n-1}.$$

A proof of this can be found in Appendix I pg 123-124.

Example 3.4 Find $f'(x)$ if

$$f(x) = 2x^4 - x^2 + 1.$$

Solution 3.4 We have, by theorem 3.2,

$$\begin{aligned} f'(x) &= 2(4)x^{4-1} - 2x^{2-1} + 0 \\ &= \boxed{8x^3 - 2x}. \end{aligned}$$

Theorem § 3.3: The product rule Lets say you have a product of two functions. I.e., if

$$f(x) = g(x)h(x),$$

then

$$f'(x) = g'(x)h(x) + g(x)h'(x).$$

Example 3.5 Find $f'(x)$ if

$$f(x) = x^2 \ln x.$$

Solution 3.5 From last section, the derivative of $\ln x$ is $\frac{1}{x}$ so $g'(x) = 2x, h'(x) = \frac{1}{x}$ and we get

$$\begin{aligned} f'(x) &= g'(x)h(x) + g(x)h'(x) \\ &= 2x \ln x + x^2(\frac{1}{x}) \\ &= \boxed{x(2\ln x + 1)}. \end{aligned}$$

Theorem § 3.4: The quotient rule Let us say you have a rational function. So if

$$f(x) = \frac{g(x)}{h(x)},$$

then

$$f'(x) = \frac{h(x)g'(x) - g(x)h'(x)}{(h(x))^2} = \frac{\text{lo d hi - hi d lo}}{(\text{lo})^2}.$$

Here, lo is the bottom function, hi is the top and a d in front means derivative of. Just a cool little way I always used to remember this rule.

Example 3.6 Find $f'(x)$ if

$$f(x) = \frac{\ln x}{x}.$$

Solution 3.6 We have by the quotient rule, $g(x) = \ln x, h(x) = x$ so $g'(x) = \frac{1}{x}, h'(x) = 1$. Thus

$$\begin{aligned} f'(x) &= \frac{\frac{1}{\cancel{x}}(\cancel{x}) - \ln x}{(x)^2} \\ &= \boxed{\frac{1 - \ln x}{x^2}}. \end{aligned}$$

We could have also for Example 3.6 converted the function using our exponent rules into

$$x^{-1} \ln x$$

and invoked the product rule.

Theorem § 3.5: The chain rule Let's say you have a function inside a function. Then you have if

$$f(x) = g(h(x)),$$

then

$$f'(x) = h'(x)g'(h(x)).$$

A formal proof can be found in Appendix I pg 124. (Formally, g differentiable at x_0, f differentiable at $g(x_0)$..)

Example 3.7 Find $f'(x)$ if

$$f(x) = (x^2 + 2)^8.$$

Solution 3.7 Here our $h(x) = x^2 + 2$ and our $g(x) = x^8$. Thus $h'(x) = 2x$ and we have

$$\begin{aligned} f'(x) &= h'(x)g(h(x)) \\ &= (2x)(8)(x^2 + 2)^{8-1} \\ &= \boxed{16x(x^2 + 2)^7}. \end{aligned}$$

Theorem § 3.6: The rules for derivatives Let f, g be differentiable functions. Then if $b \in \mathbb{R}$, one has

1. $\frac{d}{dx}[bf(x)] = b\frac{d}{dx}[f(x)]$.
2. $\frac{d}{dx}[f(x) \pm g(x)] = \frac{d}{dx}[f(x)] \pm \frac{d}{dx}[g(x)]$.

Theorem § 3.7: Derivatives of transcendental: The following are some useful tricks:

1. If $f(x) = \sin x$, then $f'(x) = \cos x$.
2. If $f(x) = \cos x$, then $f'(x) = -\sin x$.
3. If $f(x) = e^{g(x)}$, then $f'(x) = g'(x)e^{g(x)}$.
4. If $f(x) = \ln(g(x))$, then $f'(x) = \frac{g'(x)}{g(x)}$.
5. Expanding the formulas in (1) using the chain rule, if $f(x) = \sin(g(x))$, then $f'(x) = g'(x)\cos(g(x))$. And similarly, if $f(x) = \cos(g(x))$, then $f'(x) = -g'(x)\sin(g(x))$.

Theorem § 3.8: Derivatives of inverse trig functions We have

1. If $f(x) = \sin^{-1}(g(x))$, then $f'(x) = \frac{g'(x)}{\sqrt{1-(g(x))^2}}$.
2. If $f(x) = \cos^{-1}(g(x))$, then $f'(x) = -\frac{g'(x)}{\sqrt{1-(g(x))^2}}$.
3. If $f(x) = \tan^{-1}(g(x))$, then $f'(x) = \frac{g'(x)}{1+(g(x))^2}$.
4. If $f(x) = \csc^{-1}(g(x))$, then $f'(x) = -\frac{g'(x)}{|g(x)|\sqrt{(g(x))^2-1}}$.
5. If $f(x) = \sec^{-1}(g(x))$, then $f'(x) = \frac{g'(x)}{|g(x)|\sqrt{(g(x))^2-1}}$.
6. If $f(x) = \cot^{-1}(g(x))$, then $f'(x) = -\frac{g'(x)}{1+(g(x))^2}$.

Example 3.8(a) Find $f'(x)$ if

$$f(x) = e^{x^2}.$$

Solution 3.8(a) Here we have the situation of Theorem 3.7.3 where $g(x) = x^2$ so then $g'(x) = 2x$ by the power rule and

$$\begin{aligned} f'(x) &= \frac{d}{dx}[e^{x^2}] \\ &= \frac{d}{dx}[x^2]e^{x^2} \\ &= \boxed{2xe^{x^2}}. \end{aligned}$$

Example 3.8(b) Find $f'(x)$ if

$$f'(x) = \ln(4x^4).$$

Solution 3.8(b) Here we are in situation of Theorem 3.7.4 where $g(x) = 4x^4$. We then have

$$\begin{aligned} f'(x) &= \frac{g'(x)}{g(x)} \\ &= \frac{16x^3}{4x^4} \\ &= \boxed{\frac{4}{x}}. \end{aligned}$$

A general rule is if your natural log has one term, i.e., if

$$f(x) = \ln(ax^n),$$

then

$$\boxed{f'(x) = \frac{n}{x}}.$$

Example 3.9(a) Find $f'(x)$ if

$$f(x) = \csc^{-1}\left(\frac{1}{2}x^2\right).$$

Solution 3.9(a) Here we have Theorem 3.8.4 where $g(x) = \frac{1}{2}x^2$ thus $g'(x) = x$. And so

$$\begin{aligned} f'(x) &= -\frac{g'(x)}{|g(x)|\sqrt{(g(x))^2 - 1}} && \text{Theorem 3.8} \\ &= -\frac{x}{|\frac{1}{2}x^2|\sqrt{(\frac{1}{2}x^2)^2 - 1}} && \text{plugging in} \\ &= -\frac{x}{\frac{1}{2}x^2\sqrt{\frac{1}{4}x^4 - 1}} && \text{since } x^2 > 0 \text{ we drop the absolute value} \\ &= -\frac{2}{x\sqrt{\frac{x^4-4}{4}}} && \text{combine fractions} \\ &= \boxed{-\frac{4}{x\sqrt{x^4 - 4}}} && \text{simplifying.} \end{aligned}$$

Example 3.9(b) Find $f'(x)$ if

$$f(x) = \tan x.$$

Solution 3.9(b) Here we rewrite tangent as

$$\tan x = \frac{\sin x}{\cos x}.$$

And then apply the quotient rule:

$$\begin{aligned} f'(x) &= \frac{\frac{d}{dx}[\sin x]\cos x - \frac{d}{dx}[\cos x]\sin x}{(\cos x)^2} && \text{By the quotient rule} \\ &= \frac{\cos^2 x + \sin^2 x}{\cos^2 x} && \text{Theorem 3.7.1} \\ &= \frac{1}{\cos^2 x} && \text{by chapter 1 trig review} \\ &= \boxed{\sec^2 x} && \text{by chapter 1 trig review.} \end{aligned}$$

Now we write equations of tangent lines.

Example 3.10 Write the equation of the line tangent to

$$f(x) = x^3 + x - 1$$

at $x = 2$.

Solution 3.10 First we find y. We have

$$f(2) = 2^3 + 2 - 1 = 8 + 2 - 1 = 9.$$

And the derivative is

$$f'(x) = 3x^2 + 1.$$

So the slope of our line is

$$f'(2) = 3(2)^2 + 1 = 13.$$

So then the y–intercept is

$$9 = (13)(2) + b \Rightarrow b = -17.$$

so the equation of our line is

$$\boxed{y = 13x - 17}.$$

Below is an image of the curve in blue and the tangent line in red, and the point of intersection at $(2, 9)$:

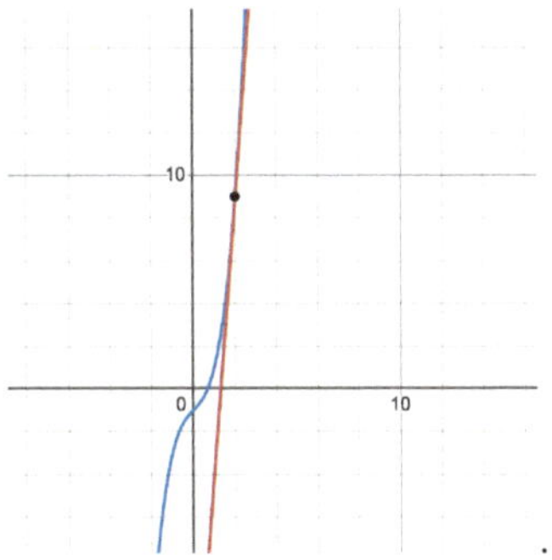

Example 3.11 Find the equation of the line tangent to

$$f(x) = e^{x+1}$$

at $x = 1$.

Solution 3.11 For the $y-$value, $f(1) = e^{1+1} = e^2$. And the derivative is e^{x+1} so at $x = 1$ we get $f'(1) = e^2$ so slope is e^2 then for $y-$intercept

$$e^2 = (e^2)(1) + b \Rightarrow b = 0$$

thus the tangent line is

$$\boxed{y = e^2 x}.$$

Below in blue is $f(x)$, red is the line tangent to $f(x)$ at $x = 1$, $y = e^2 x$ and the point of intersection $(2, e^2)$:

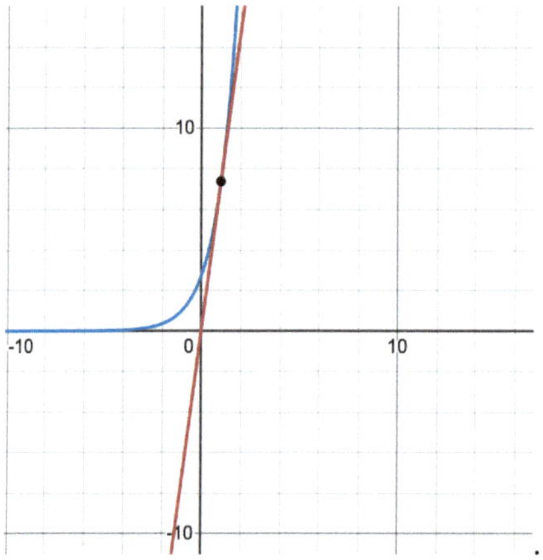

>§ 3.2: Exercises

1. Find f' if
$$f(x) = \frac{1}{4}x^4 + \frac{1}{3}x^3 + \frac{1}{2}x^2 + x + \pi.$$

2. Find f' if
$$f(x) = \sqrt{x} + \frac{1}{x^4}.$$

3. Find f' if
$$f(x) = e^x x^2.$$

4. Find f' if
$$ff(x) = \ln(10x^9).$$

5. Find f' if
$$f(x) = \sin(x^2).$$

6. Find f' if
$$f(x) = 2\tan(\frac{1}{2}x^2).$$

7. Find f' if
$$f(x) = \sin^{-1}(2x^3).$$

8. Find f' if
$$f(x) = \frac{x^2}{e^x}.$$

9. Find f' if
$$f(x) = 2(2x^2 + 1)^8.$$

10. Find f' if
$$f(x) = \sqrt{3x^2 + 1}.$$

11. Find f' if
$$f(x) = \sin(2\sqrt{x}).$$

12. Find f' if
$$f(x) = x^{\frac{3}{2}} - x + 1.$$

13. Find the equation of the line tangent to
$$f(x) = x^2 + 4x - 1$$
at $x = 1$.

14. Find the equation of the line tangent to
$$f(x) = e^{x^2}$$
at $x = 1$.

15. Find the equation of the line tangent to

$$f(x) = \sin 2x$$

at $x = \frac{1}{4}$.

>§ 3.3: Implicit differentiation & Related rates

What would happen if we were differentiating two variables at once? So far we have seen only an x and we apply $\frac{d}{dx}$ but if we have say

$$F(x, y)$$

some function, in order to find the *differential* $\frac{dy}{dx} = y'$, we must attach a $d\_\_\_$ to let the reader know which variable you were differentiating. And we divide all terms by dx to get $\frac{dy}{dx}$ which is read "*the derivative of y with respect to x*" and this will come in handy when dealing with related rates problems which include a dimensional change as well as a time change. If you have x, y being multiplied you must invoke the product rule and treat one as a constant and the other as a variable.

Example 3.12(a) Find $\frac{dy}{dx}$ in

$$x^2 - y^2 + xy = 4.$$

Example 3.12(a) We have

$$\begin{aligned}
&d(x^2 - y^2 + xy = 4) \Rightarrow 2xdx - 2ydy + ydx + xdy = d(4) \\
&\Rightarrow 2x - 2y\frac{dy}{dx} + y + x\frac{dy}{dx} = 0 \\
&\Rightarrow 2x + y = \frac{dy}{dx}(2y - x) \\
&\Rightarrow \boxed{\frac{dy}{dx} = \frac{2x + y}{2y - x}}.
\end{aligned}$$

Example 3.12(b) Find the equation of the line tangent to the curve in 3.12(a) at the point (1,1).

Solution 3.12(b) At (1,1) the slope is

$$\frac{dy}{dx} = \frac{2(1) + 1}{2(1) - 1} = 3.$$

And so the $y-$intercept is

$$1 = (3)(1) + b \Rightarrow b = -2$$

so the lines equation is

$$\boxed{y = 3x - 2}.$$

Related rates are world problems in which we find the rate at which a dimension changes with respect to time. So imagine we have:

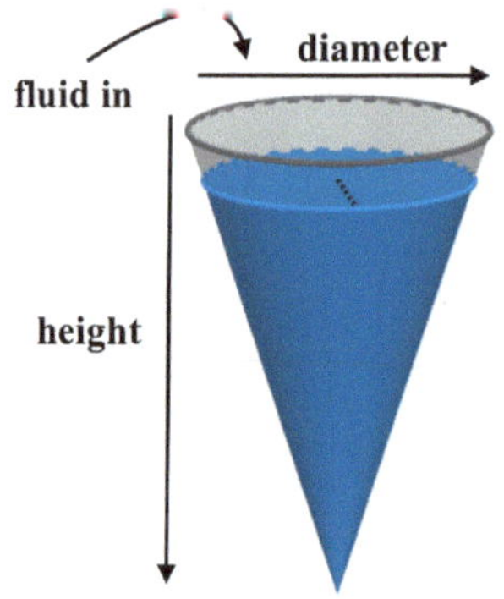

.

Example 3.13(a) Imagine a cone whose height is 2 feet and radius is 4 feet is being filling up at a rate of 2 cubic inches per second. At what rate is the height of the water increasing?

Solution 3.13(a) We know

$$V = \frac{1}{3}\pi r^2 h.$$

If $h = 2$ and $r = 4$ then $r = 2h$ so we have

$$V = \frac{1}{3}\pi(2h)^2 h = \frac{4}{3}\pi h^3.$$

And applying implicitly $\frac{d}{dt}$ we get

$$\frac{d}{dx}[V = \frac{4}{3}\pi h^3] \Rightarrow \frac{dV}{dt} = 4\pi h^2 \frac{dh}{dt}.$$

I.e.,

$$V' = 4\pi h^2 h'.$$

Usually in this situation, you have say 3 unknowns and 2 knowns and so we know $\frac{dV}{dt} = 2$ and $h = 2$ so

$$2 = 4\pi(2)^2 \frac{dh}{dt} \Rightarrow \boxed{\frac{dh}{dt} = \frac{1}{8\pi}\text{in}^3/\text{sec}}.$$

When dealing with a ladder sliding down a wall, as the photo below shows:

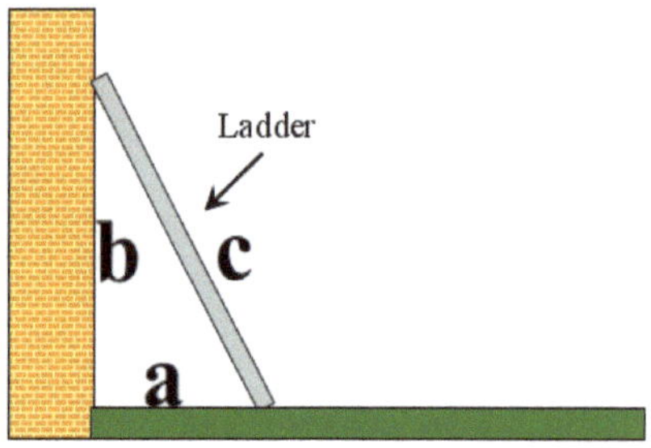

.

we start with

$$a^2 + b^2 = c^2$$

and $\frac{d}{dt}$ of this gives us

$$2aa' + 2bb' = 0$$

or just

$$\boxed{aa' + bb' = 0}$$

because $c' = 0$ since c represents the latter and here the derivative is zero as the height never changes and a, b are sides from top to floor and floor to bottom and a', b' rates at which top going down an bottom slipping away. I.e.,

$$a' = \frac{da}{dt}, b' = \frac{db}{dt}.$$

Example 3.13(b) A 5 foot latter is sliding down at a rate of 2 inches per second. At the moment the top is 4 feet high what rate is the bottom slipping away at?

Solution 3.13(b) We know that to find side a,

$$a^2 + 4^4 = 5^2 \Rightarrow a = 5.$$

And $b = 4, b' = -2$ so again, here we have 4 unknowns and 3 knowns thus since $aa' + bb' = 0$,

$$3a' - (2)(4) = 0 \Rightarrow a' = \boxed{\frac{8}{3}}$$

inches per second.

Example 3.13(c) An air balloon is deflating at a rate of 4 inches cubed per second. At the instant the diameter is 6 inches, find the rate at which the radius is changing.

Solution 3.13(c) The formula is

$$V = \frac{4}{3}\pi r^3$$

and applying $\frac{d}{dt}$ we obtain

$$V' = 4\pi r^2 r'.$$

Here, it is implied that $V' = \frac{dV}{dt}$ and $r' = \frac{dr}{dt}$. We then have an equation with 3 unknowns, 2 of which are given so we can find the third: We have that $V' = -4, r = 3$ since $d = 6$. So we are on the hunt to find r', or $\frac{dr}{dt}$. Thus

$$\begin{aligned}
&\Rightarrow -4 = 4\pi(3)^2 r' \\
&\Rightarrow -1 = 9\pi r' \\
&\Rightarrow \boxed{r' = -\frac{1}{9\pi}}.
\end{aligned}$$

So radius is decreasing at $\frac{1}{9\pi}$ inches per second.

For a brief moment we discuss higher order derivatives, i.e., we write

$$f^{(n)}(x)$$

to denote the $n-$th derivative of f with respect to x. So $f'(x)$ is said f *prime of* x. And $f^{''}(x), f^{'''}(x)$ are often used to denote the second and third derivatives respectively. If a function's derivative of n order exists, you may differentiate n times.

Example 3.14(a) Find $f^{'''}(x)$ if

$$f(x) = x^4 + x^3.$$

Solution 3.14(a) We have that

$$f'(x) = 4x^3 + 3x^2.$$

And so

$$f^{''}(x) = 12x^2 + 6x.$$

Thus the third derivative is

$$f^{'''}(x) = 24x + 6.$$

Example 3.14(b) Find $f^{''}(x)$ if

$$f(x) = x \ln x.$$

Solution 3.14(b) Note, using power rule,

$$f'(x) = \ln x + 1.$$

And once more,

$$f^{''}(x) = \frac{1}{x}.$$

In physics, the *velocity function* is the derivative of the position function. And the *acceleration* is the derivative of velocity. for the $n-$the derivative of f with respect to x. A few notes on continuity and differentiability, if a function is differentiable, at a point, it is certainly continuous at this point but vis versa is not true. Also if a function has one derivative, all of them need not exist. Furthermore, a continuous function need not have continuous derivative.

>§ 3.3: Exercises

1. Find $\frac{dy}{dx}$ if
$$x^2 - \frac{1}{2}y^2 - xy = 14.$$

2. Find $\frac{dy}{dx}$ if
$$x^2 - e^y = 14.$$

3. Find $\frac{dy}{dx}$ if
$$\frac{1}{2}x^2 + \ln y = 2.$$

4. If air is going into a balloon at a rate of 6 inches cubed per second, find the rate the radius is changing the instant the diameter is 12 inches.

5. A 13 food latter is sliding down a wall at a rate of 5 inches per second. At the moment the top is 5 inches from the floor at what rate is the bottom slipping away?

6. A cone has water leaking from it at a rate of 6 inches cubed per second. The height is twice the radius. When the height is 3 inches, at what rate is it changing?

7. Find $f^{(4)}(x)$ if
$$f(x) = \sin(2x).$$

8. Find $f^{''}(x)$ if
$$f(x) = xe^x.$$

9. Find $f^{'''}(x)$ if
$$f(x) = x^4 + x^3 + x^2 + x + 1.$$

>§ 3.4: Derivatives of inverse functions

Recall, that two functions f, g are inverses of one another if

$$f(g(x)) = x = g(f(x)).$$

Two facts which we will not present as Theorems which are:

1. If a function is continuous, so is its inverse.
2. If a function is differentiable, then so is its inverse given than $f' \neq 0$ at the point of differentiability.

Theorem § 3.9: Derivative of an inverse Let f be a differentiable function. If g is the inverse of f, then for all x values except for the one which $f'(g(x)) \neq 0$, one has

$$g'(x) = \frac{1}{f'(g(x))}.$$

Example 3.15 If

$$f(x) = \frac{1}{4}x^3 + x - 1$$

and $f(2) = 3$, without computing the inverse of f, find $f^{-1}(3)$ and $(f^{-1})'(3)$.

Solution 3.15 Since $f(2) = 3$, then the inverse of 3 is just 2!. For the derivative of its inverse, we must first find $f'(f^{-1}(3))$. But this is $f'(2)$ and thus

$$f'(x) = \frac{3}{4}x^2 + 1$$

So at $x = 2$ we get $f'(2) = \frac{3}{4}(2)^2 + 1 = 4$ thus the derivative of the inverse at 3 is

$$\boxed{\frac{1}{4}}.$$

Example 3.16 If

$$f(x) = \frac{1}{2}x^2 - x + 1,$$

and $f(0) = 1$, without computing the inverse of f, find $f^{-1}(1)$ and $(f^{-1})'(1)$.

Solution 3.16 Since $f(0) = 1$, then the inverse of 1 is just 0! And $f'(f^{-1}(1)) = f'(0)$ but $f'(x) = 2x - 1$ so for $x = 0$ we get

$$(f^{-1})'(1) = \frac{1}{f'(f^{-1}(1))} = \frac{1}{2(0) - 1} = \boxed{-1}.$$

Example 3.17 If

$$f(x) = 2e^x$$

find $(f^{-1})'(2e)$.

Solution 3.17 So this is

$$\frac{1}{f'(f^{-1}(2e))}.$$

But $f^{-1}(2e) = 1$ thus $f'(1) = 2e$ thus the answer is

$$\boxed{\frac{1}{2e}}.$$

§ 3.4: Exercises

1. If
$$f(x) = x^2 - x + 1,$$
and $f(2) = 3$, find $f^{-1}(3)$ and $(f^{-1})'(3)$ without computing the inverse.

2. If
$$f(x) = x^3 - 6x,$$
and $f(1) = -5$, find $f^{-1}(-5)$ and $(f^{-1})'(-5)$ without computing the inverse.

3. If
$$f(x) = \ln x,$$
and $f(e) = 1$, find $f^{-1}(1)$ and $(f^{-1})'(1)$ without computing the inverse.

4. If
$$f(x) = \sqrt{x} + x,$$
and $f(4) = 6$, find $f^{-1}(6)$ and $(f^{-1})'(6)$ without computing the inverse.

5. Find the inverse of
$$f(x) = \frac{2x+1}{3x-2}.$$

>§ 4. Applications of differentiation

>§ **4.1: Extrema on an interval** We now move onto a topic of major discussion. That is, we now would like to see what highest and lowest points a funtions values take on.

Definition 4.1 If f is defined on an interval I, we say $f(x)$ *has a maximum at* $x = x_0$ if

$$f(x_0) \geq f(x)$$

for all $x \in I$. Similarly, we say *a function* $f(x)$ *has a minimum at* $x = x_0$ *on some interval* I if

$$f(x_0) \leq f(x)$$

for every $x \in I$. These points are also said to be the *absolute extrema points* of f.

Theorem § 4.1: The extreme value theorem All continuous functions defined on $[a, b]$ attain a minimum and maximum point.

A proof of this will be given in Appendix I pg 124-125.

Definition 4.2 A *critical number,* x_0 is a number for which either

$$f'(x_0) = 0$$

or f is not differentiable at x_0.

Theorem § 4.2: How to find min max If

$$x_1, x_2, ..., x_n \in (a, b)$$

are points with $f'(x_j) = 0$ for $1 \leq j \leq n$, then the min is

$$\min\{f(x_1), f(x_2), ..., f(x_n), f(a), f(b)\}.$$

And the max is

$$\max\{f(x_1), f(x_2), ..., f(x_n), f(a), f(b)\}.$$

Definition 4.3 These min and max points are referred to as *absolute extrema* or *absolute min or max.*

Example 4.1 Find the extrema of

$$f(x) = \frac{1}{3}x^3 - \frac{3}{2}x^2 + 2,$$

on $[-1, 4]$.

Solution 4.1 First we must find critical values. And

$$f'(x) = x^2 - 3x.$$

So

$$
\begin{aligned}
f'(x) &= 0 \\
&\Rightarrow x^2 - 3x = 0 && \text{definition of } f'(x) \\
&\Rightarrow (x)(x-3) = 0 && \text{factoring} \\
&\Rightarrow (x) = 0 \text{ or } (x-3) = 0 && \text{in } \mathbb{R}, \text{ if } ab = 0,\ a = 0 \text{ or } b = 0 \\
&\Rightarrow x = 0 \text{ or } x = 3 && \text{solve for } x \text{ in } x-1=0, x-2=0.
\end{aligned}
$$

So the derivative is zero when $x = 0, 3$ so we must test $x = -1, 0, 3, 4$. At $x = -1$:

$$f(-1) = \frac{1}{6}.$$

Then at $x = 0$:

$$f(0) = 2.$$

Then at $x = 3$,

$$f(3) = -\frac{5}{2}.$$

And at $x = 4$:

$$f(4) = -\frac{2}{3}$$

So $\boxed{(0,2)}$ is the $\boxed{\text{max}}$ and $\boxed{(3, -\frac{5}{2})}$ is the $\boxed{\text{min}}$. Below is a graph on the interval, between the red, of the function and its extrema:

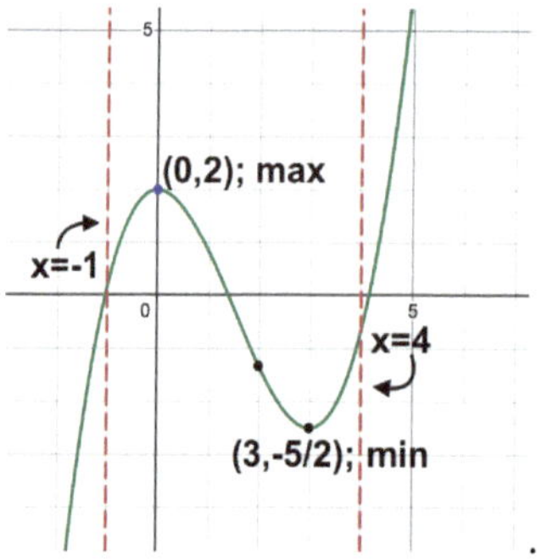

Example 4.2 Find the extrema of

$$f(x) = e^x - x$$

on $[-2, 1]$.

Solution 4.2 For critical numbers,

$$f'(x) = e^x - 1$$

which is zero at $x = 0$. This is because we have

$$\begin{aligned} f'(x) &= 0 \\ &\Rightarrow e^x - 1 = 0 && \text{definition of } f'(x) \\ &\Rightarrow e^x = 1 && \text{add 1 to both sides} \\ &\Rightarrow \ln(e^x) = \ln(1) && \text{to undo } e^x\text{, we take natural logs} \\ &\Rightarrow x = 0 && \text{because } \ln 1 = 0, \ln(e^x) = x. \end{aligned}$$

So we test $x = -2, 0, 1$. At $x = -2$,

$$f(-2) = 2 + e^{-2}.$$

at $x = 0$, we get

$$f(0) = e^0 = 0 = 1.$$

And lastly, at $x = 1$, we get

$$f(1) = e^1 - 1 = e - 1.$$

So we have $\boxed{\text{max}}$ at $\boxed{(-2, 2 + e^{-2})}$ and $\boxed{\text{min}}$ at $\boxed{(0, 1)}$.

If the interval for us is not closed, then this may fail. I.e., $f(x) = x$ has no min or max on $(0, 1)$. If you say .1 is min, I say .01, you say .001, I say .0001, so on and so forth. And similarly for a number close to 1. This is because we are not including the boundary points.

Next, we discuss how to determine where a function is increasing or decreasing.

Theorem § 4.3: Increase/Decrease intervals Let f be a function with critical numbers

$$x_1, x_2, ..., x_n \in (a, b).$$

Then to see if f is increasing on say (x_j, x_{j+1}), pick a point $p \in (x_j, x_{j+1})$, and then

1. If $f'(p) > 0$, f is increasing on (x_j, x_{j+1}).
2. If $f'(p) < 0$, f is decreasing on (x_j, x_{j+1}).

Example 4.3(a) Find which intervals

$$f(x) = x^3 - x^2$$

increases or decreases.

Solution 4.3(a) First we find f':

$$f'(x) = 3x^2 - 2x.$$

This is zero t $x = 0, \frac{2}{3}$. So we test $x = -1, \frac{1}{3}, 1$. For $x = -1$,

$$f'(-1) = 5 > 0$$

so $\boxed{\text{increasing on}}$ $\boxed{(-\infty, 0)}$. At $x = \frac{1}{3}$,

$$f'(\frac{1}{3}) = -\frac{1}{3} < 0$$

$\boxed{\text{decreasing}}$ on $\boxed{(0, \frac{2}{3})}$. Lastly, at $x = 1$, we get $f'(1) = 1 > 0$, so $\boxed{\text{increasing}}$ on $\boxed{(\frac{2}{3}, \infty)}$. This can be seen below:

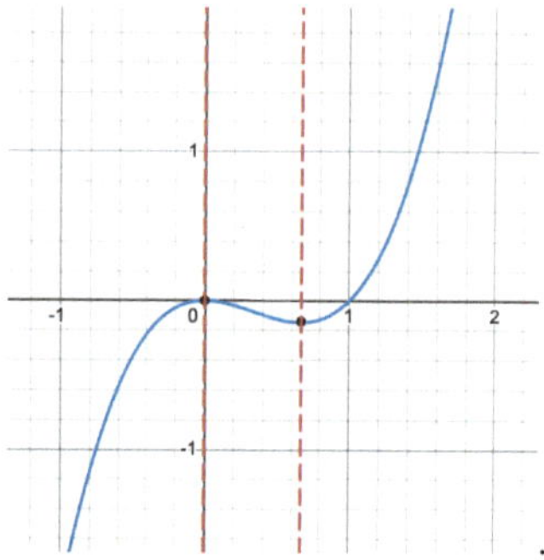

Example 4.3(a) Find the intervals of increase and decrease for

$$f(x) = x^2 - 2x.$$

Solution 4.3(a) First we find critical numbers, so

$$\begin{aligned} &\Rightarrow f'(x) = 0 \\ &\Rightarrow 2x - 2 = 0 \\ &\Rightarrow 2x = 2 \\ &\Rightarrow x = 1. \end{aligned}$$

So we test before and after 1 so say $x = 0, 2$ we test into the derivative. At $x = 0$,

$$f'(0) = -2 < 0$$

thus we are $\boxed{\text{decreasing}}$ on $(-\infty, 1)$. And at $x = 2$:

$$f'(2) = 2 > 0$$

so we $\boxed{\text{increase}}$ on $\boxed{(1, \infty)}$. As seen below:

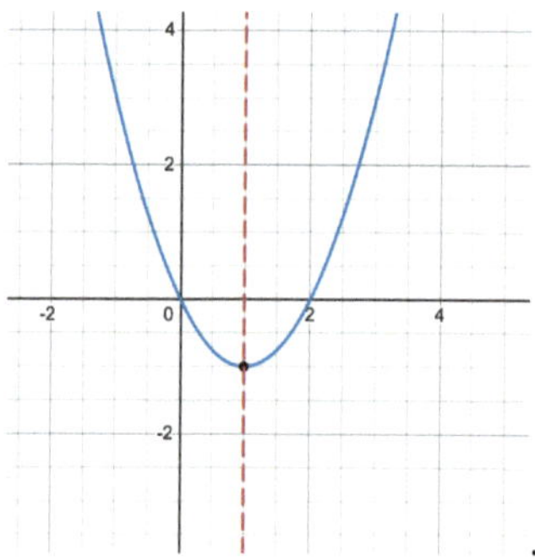

§ 4.1 Exercises

1. Find the min and max of
$$f(x) = 2x^2 - 4x$$
on $[0, 2]$.

2. Find the min and max of
$$f(x) = \frac{1}{3}x^3 - \frac{1}{2}x^2 - \frac{1}{2}.$$
on $[-1, 2]$.

3. Find the min and max of
$$f(x) = e^x - 2x$$
on $[0, 1]$.

4. Find the intervals of increase and decrease for (1).

5. Find the intervals of increase and decrease for (2).

6. Find the intervals of increase and decrease for (3).

>§ 4.2: Rolle's, Mean Value, & Intermediate Value Theorems

Theorem § 4.4: Rolle's Theorem Let f be continuous on $[a, b]$ and differentiable on (a, b). If

$$f(a) = f(b),$$

then there exists some $x_0 \in (a, b)$ such that

$$f'(x_0) = 0.$$

Theorem § 4.5: The Mean Value Theorem Let f be continuous on $[a, b]$ and differentiable on (a, b). Then there exists some $x_0 \in (a, b)$ such that

$$f'(x_0) = \frac{f(b) - f(a)}{b - a}.$$

Theorem § 4.5: The Intermediate Value Theorem Let f be continuous on $[a, b]$. If there is some y with

$$\min\{f(a), f(b)\} < y < \max\{f(a), f(b)\},$$

then there is some $x_0 \in (a, b)$ such that

$$f(x_0) = y.$$

This last theorem can be used for the existence of zeroes as in this next example:

Example 4.4 If

$$f(x) = \frac{1}{3}x^3 - x - \frac{1}{2},$$

show using the intermediate value theorem that it has a zero on $(0, 2)$.

Solution 4.4 We compute $f(0), f(2)$:

$$f(0) = -\frac{1}{2}.$$

And

$$f(2) = \frac{1}{3}2^3 - 2 - \frac{1}{2} = \frac{1}{6}$$

and we know

$$f(0) = -\frac{1}{2} < 0 < \frac{1}{6} = f(2),$$

so then, by intermediate value theorem, there must be a $x_0 \in (0, 2)$ such that

$$f(x_0) = 0.$$

Example 4.5 Show the following satisfies Rolle's theorem and if so find the x_0.

$$f(x) = x^2 + 1$$

on $[-2,2]$.

Solution 4.5 We have to satisfy the theorem so

$$\begin{aligned} f(a) &= f(-2) \\ &= (-2)^2+1 \\ &= 5 \\ &= (2)^2+1 \\ &= f(2) \\ &= f(b). \end{aligned}$$

So then to find the $x_0 \in (-2,2)$,

$$f'(x_0) = 2x_0$$

and equals zero when $\boxed{x_0 = 0}$.

Example 4.6 Use the mean value theorem to find the $x_0 \in (0,2)$ for

$$f(x) = x^2 + x - 2.$$

Solution 4.6 So we must find $f(0), f(2)$. So $f(0) = -2, f(2) = 4$ and $f'(x) = 2x+1$ so

$$\begin{aligned} f'(x_0) &= \frac{f(b)-f(a)}{b-a} \\ &= \frac{6}{4} \\ &= \frac{3}{2}. \end{aligned}$$

So

$$f'(x_0) = 2x - \frac{1}{2} = 0 \Rightarrow \boxed{x = -\frac{1}{4}}.$$

So in Mean value theorem you are finding an x_0 so that $f'(x_0)$ is $\frac{f(b_f(a)}{b-a}$ and in Rolle's we say wait a minute what if $f(b = f(a)$, then in Rolle's we find x_0 in the interval such that $f'(x_0) = 0$. Intermediate value guarantees us the existence of a zero in the event that $f(a), f(b)$ take on opposite signes because only then is the number between them a 0!! So you know there is a number between a, b that gets hit to zero! By intermediate value!!

>§ 4.2 Exercises

1. Determine if Rolle's theorem applies to
$$f(x) = (x-1)^2 + 2$$
on $[-2, 2]$. If so find the $x_0 \in (-2, 2)$ such that
$$f'(x_0) = 0.$$

2. Determine if Rolle's theorem applies to
$$f(x) = (x-2)^2 - 1$$
on $[1, 3]$. If so find the $x_0 \in (1, 3)$ such that
$$f'(x_0) = 0.$$

3. Use the Mean Value theorem to find $x_0 \in (-1, 1)$ such that
$$f'(x_0) = \frac{f(b) - b(a)}{b - a}$$
where $a = -1, b = 1$ for
$$f(x) = (x-2)^2 + 4.$$

4. Use the Mean Value theorem to find $x_0 \in (0, 3)$ such that
$$f'(x_0) = \frac{f(b) - b(a)}{b - a}$$
where $a = 0, b = 3$
$$f(x) = x^2 - 2x.$$

5. Use the Intermediate Value Theorem to show
$$f(x) = x^3 - 2x + 1$$
has a zero on $[-2, 2]$ call it x_0.

6. Use the Intermediate Value Theorem to show
$$f(x) = 2x^2 - 4x + 1$$
has a zero on $[-1, 1]$ call it x_0.

>§ 4.3: First & Second derivative tests

We have the following test to determine *relative max or relative min*:

Theorem § 4.6: The First derivative test Let x_0 be a critical number of f with $x_0 \in (a, b)$ such that f is continuous on (a, b) and differentiable except for possibly at x_0, then

1. If $f'(x_0)$ goes from negative to positive, then $(x_0, f(x_0))$ is a *relative minimum*.
2. If $f'(x_0)$ goes from positive to negative , then $(x_0, f(x_0))$ is a *relative maximum* .
3. If $f'(x_0)$ is either both negative or both positive before and after, the point is neither minimum nor maximum.

Example 4.7 Given

$$f(x) = x^3 - x^2,$$

find the relative extrema.

Solution 4.7 First we find f':

$$f'(x) = 3x^2 - 2x.$$

Thus

$$\begin{aligned} f'(x) &= 0 \\ &\Rightarrow 3x^2 - 2x = 0 \\ &\Rightarrow x(3x - 2) = 0 \\ &\Rightarrow x = 0 \text{ or } 3x - 2 = 0. \\ &\Rightarrow x = 0 \text{ or } x = \frac{2}{3}. \end{aligned}$$

So for relative extrema, for $x = 0$,

$$f'(-1) = 5,$$
$$f(\frac{1}{3}) = -\frac{1}{3}$$

so $\boxed{(0, 0)}$ is a $\boxed{\text{relative maximum}}$ while

$$f(\frac{1}{3}) = -\frac{1}{3}$$

and

$$f(1) = 1$$

so we go from negative to positive so

$$\boxed{\left(\frac{2}{3}, -\frac{4}{27}\right)},$$

which is a $\boxed{\text{relative minimum}}$.

A reminder that not all functions have relative min and max. I.e.,

$$f(x) = e^x$$

does not have a relative min nor max. On an interval it will have an absolute min and max on its end points. A big reason is that it is its own derivative. I.e., $\frac{d}{dx}[e^x] = e^x$, and

$$e^x > 0$$

for any value of x.

Definition 4.4 If $f''(x_0)$ is either undefined or zero, we say x_0 is an *infliction point of* $f(x)$. A graph is *concave up* if it is facing upwards and *concave down* if facing down. More specifically, a function f is *concave up* if f' is increasing on the interval and *concave down* if f' is decreasing on the interval. Below is a photo of the scenarios:

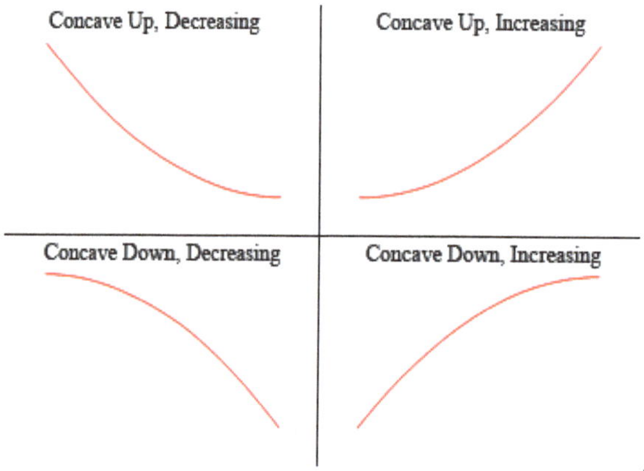

Theorem § 4.7: Test for concavity Let f be a twice differentiable function and x_0 an infliction point. If p is a point in an interval which was split up using the infliction point then

1. If $f''(p) > 0$ for all p in the interval, then f is concave up on this interval.
2. If $f''(p) < 0$ for all p in the interval, then f is concave down on this interval.

Theorem § 4.8: Second derivative test Let f be twice differentiable and suppose x_0 is a critical number for f. Then

1. If $f''(x_0) > 0$, then f has a relative minimum at $(x_0, f(x_0))$.
2. If $f''(x_0) < 0$, then f has a relative maximum at $(x_0, f(x_0))$.
3. If $f''(x_0) = 0$, the test is inconclusive, meaning it could have both or neither.

Example 4.8 Using the second derivative test, find the relative extrema for

$$f(x) = \frac{1}{3}x^3 - \frac{1}{2}x^2.$$

Solution 4.8 We know

$$f'(x) = x^2 - x.$$

And so the critical numbers are $x = 0, 1$. So for $x = 0$:

$$f''(x) = 2x - 1$$

and so

$$f''(0) = -1 < 0$$

so $\boxed{(0,0)}$ is a $\boxed{\text{relative maximum}}$. And

$$f''(1) = 1 > 0$$

so we have a $\boxed{\text{relative minimum}}$ at $\boxed{(1, -\frac{1}{6})}$.

Theorem § 4.9: Test for concavity using second derivative For a function f and infliction points

$$x_1, x_2, ..., x_n$$

if $p \in (x_j, x_{j+1})$, then

1. If $f''(p) > 0$, then f is concave up on the interval p is in.
2. If $f''(p) < 0$, then f is concave down on the interval p is in.

Example 4.9 Test for intervals of concavity for

$$f(x) = x^3 - x^2.$$

Solution 4.9 We know

$$f'(x) = 3x^2 - 2x.$$

Thus

$$f''(x) = 6x - 2.$$

Thus the infliction point is just $\frac{1}{3}$. So we test before this at $x = 0$:

$$f''(0) = -2 < 0$$

so we are $\boxed{\text{concave down}}$ on $\boxed{(-\infty, \frac{1}{3})}$. And for after, we test at $x = 1$:

$$f''(1) = 4 > 0$$

so we are $\boxed{\text{concave up}}$ on $\boxed{(\frac{1}{3}, \infty)}$.

>§ 4.3 Exercises

1. Find relative extrema using first and second derivatives tests:

$$f(x) = 2x^2 - x + 1.$$

2. Find relative extrema using first and second derivatives tests:

$$f(x) = x^3 - \frac{1}{2}x^2.$$

3. Find relative extrema using first and second derivatives tests:

$$f(x) = x^2 - 2x + 1.$$

4. Find relative extrema using first and second derivatives tests:

$$f(x) = x \ln x.$$

5. Find relative extrema using first and second derivatives tests:

$$f(x) = xe^x.$$

6. Determine the concavity intervals for

$$f(x) = \frac{1}{3}x^3 - \frac{1}{2}x^2.$$

7. Determine the concavity intervals for

$$f(x) = 2 \sin x.$$

8. Determine the concavity intervals in

$$f(x) = 4x^3 - x^2 + 4x.$$

9. Determine the concavity intervals in (5).

>§ 4.4: Limits @ infinity

We now move onto the topic of determining various limits at infinite. That is,

$$\lim_{x\to\infty} f(x) = L.$$

Definition 4.5 A function f has *limit* L *at* ∞ if for any $\epsilon > 0$ there is a $\delta > 0$ such that if

$$x > \delta,$$

then we still have

$$|f(x) - L| < \epsilon.$$

Similarly, we say f *has limit* L *as* $x \to -\infty$ if for any $\epsilon > 0$ there is a $\delta < 0$ such that if

$$x < \delta,$$

then we still have

$$|f(x) - L| < \epsilon.$$

Definition 4.6 If a function has a horizontal asymptotes at say $y = L$, then

$$\lim_{x\to\pm\infty} f(x) = L.$$

Theorem § 4.10: Rules for rational limits at infinity We have If $n \in \mathbb{Q}^+$, and $x_0 \in \mathbb{R}$, then

$$\lim_{x\to\infty} \frac{x_0}{x^n} = 0.$$

Furthermore, if the function is defined when $x < 0$, then

$$\lim_{x\to-\infty} \frac{x_0}{x^n} = 0.$$

Side-Note: You can split up the sum or difference of limits in the infinity case in the event they both exist and are finite, I.e.,

$$\lim_{x\to\infty} [f(x) \pm g(x)] = \lim_{x\to\infty} f(x) \pm \lim_{x\to\infty} g(x).$$

Example 4.10 Evaluate

$$\lim_{x\to\infty} (4 - \frac{5}{x^4}).$$

Solution 4.10 We have

$$\begin{aligned}
\lim_{x\to\infty} (4 - \frac{5}{x^4}) &= \lim_{x\to\infty} 4 - \lim_{x\to\infty} \frac{5}{x^4} && \text{by side-note} \\
&= 4 - 0 && \text{by theorem 4.10.1\& 2.1.1} \\
&= \boxed{4}.
\end{aligned}$$

Example 4.11(a) Find

$$\lim_{x\to\infty} \frac{2x^2+3}{3x^2+1}.$$

Solution 4.11(a) Since this function has a horizontal asymptote at $y = \frac{2}{3}$, the limit is $\boxed{\frac{2}{3}}$.

Example 4.11(b) Find

$$\lim_{x\to\infty} \frac{2x}{x^2+1}.$$

Solution 4.11(b) This has a HA at $y = 0$ so the limit is $\boxed{0}$.

Theorem § 4.11: Laws for limits of e We have

$$\lim_{x\to\infty} e^{-x} = 0,$$

and

$$\lim_{x\to-\infty} e^{x} = 0.$$

As a consequence, things like

$$\lim_{x\to\infty} \frac{4}{e^x} = 0.$$

And one has

$$\lim_{x\to\infty} \ln x = \infty,$$

but

$$\lim_{x\to\infty} \ln(1 + \frac{1}{x}) = 0.$$

Example 4.12 Evaluate

$$\lim_{x\to\infty} \frac{2x^3}{x^2+2}.$$

Solution 4.12 Here, the degree on top is greater thus

$$\lim_{x\to\infty} \frac{2x^3}{x^2+2} = \infty.$$

Example 4.13 Evaluate

$$\lim_{x\to\infty} \frac{4}{x-2.}$$

Solution 4.13 Here, the degree on top is lesser so the function has a HA at $y = 0$ thus the limit is 0. I.e.,

$$\lim_{x\to\infty} \frac{4}{x-2} = 0.$$

>§ 4.4 Exercises

1. Evaluate
$$\lim_{x\to-\infty} \frac{e^x}{5} + 2.$$

2. Evaluate
$$\lim_{x\to\infty} \frac{2x^2}{3x^2+1} + \frac{1}{3}.$$

3. Evaluate
$$\lim_{x\to\infty} \frac{x^4}{x^2+2x-4}.$$

4. Evaluate
$$\lim_{x\to\infty} \frac{5}{x^2-1}.$$

5. Evaluate
$$\lim_{x\to\infty} \frac{3}{\ln x}.$$

6. Evaluate
$$\lim_{x\to\infty} \ln(1+\frac{1}{x}) - \frac{3x}{2x+1}.$$

>§ 4.5: Curve sketching

In this section we learn how to sketch the general shape of a given curve using calculus and algebra techniques. In doing so, we first solve for the following, for a given function, say $f(x)$, we check for:

1. $x, y-$intercepts, if there are any.
2. HA,VA if there are any or slants.
3. Domain, range.
4. If there are any symmetries.
5. Increase/decrease intervals using first derivative.
6. concavity intervals using second derivative test.

Example 4.14 Sketch the graph of

$$f(x) = \frac{4}{x-2}.$$

Solution 4.14 For (1), the $x-$intercepts is where $y = 0$ which cannot happen as the numerator is never zero so no $x-$intercepts. However for $y-$intercepts:

$$\frac{4}{0-2} = -\frac{4}{2} = -2.$$

So we have a $y-$intercept at the point $(0, -2)$. As the numerators degree is lesser, the HA is at $y = 0$ and for VA, note

$$x - 2 = 0 \Rightarrow x = 2.$$

So we have a VA at $x = 2$. As for domain, its every real number for which the denominator is nonzero:

$$x - 2 \neq 0 \Rightarrow x \neq 2.$$

Thus domain is

$$\mathbb{R} \setminus \{2\} = (-\infty, 2) \cup (2, \infty).$$

As for range, we hit all $y-$values except for 0 so the range is

$$\mathbb{R} \setminus \{0\} = (-\infty, 0) \cup (0, \infty).$$

For symmetries, it is symmetric about the origin and the line $x = 2$ i.e., $(4, 2)$ reflected has a point at $(0, -2)$. For increase or decrease, the derivative (first) is

$$f'(x) = -\frac{4}{(x-2)^2}.$$

So our intervals for testing at $(-\infty, 2)$ and $(2, \infty)$. For the first intervals, pick $x = 0$,

$$f'(0) = -\frac{4}{(-2)^2} = -1 < 0$$

so we are decreasing on $(-\infty, 2)$. Similarly, for after, the derivative is negative so decreasing on both intervals. For concavity, we have

$$f''(x) = \frac{8}{(x-2)^3}.$$

So before 2, at $x = 0$:

$$f''(0) = -1 < 0$$

so concave down on $(-\infty, 2)$ and for after, test $x = 3$:

$$f''(3) = 8 > 0$$

thus concave up on $(2, \infty)$. Thus below is the graph (blue is the graph and the dotted reds are the HA,VA):

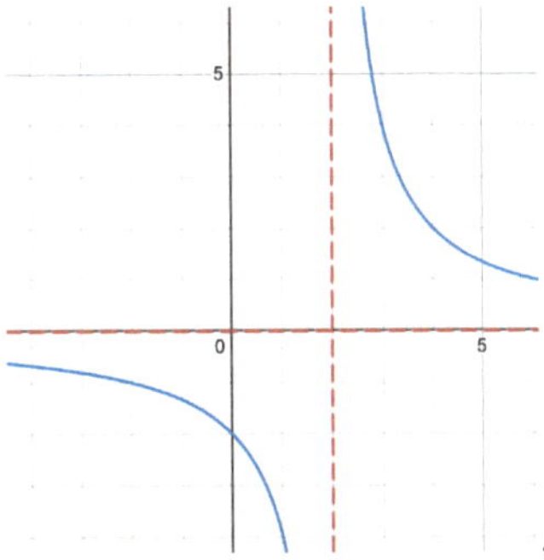

Example 4.15 Sketch the curve of

$$f(x) = \frac{1}{4}x^3 - x.$$

Solution 4.15 First off,

$$f'(x) = x^2 - 1$$

which is zero are $x = \pm 1$. And at $x = 0$,

$$f'(0) < 0$$

thus f is decreasing on $(-1, 1)$ and increasing on $(-\infty, -1) \cup (1, \infty)$. And then

$$f''(x) = 2x$$

which is zero at $x = 0$ and before zero we have $f'' < 0$ and positive after zero thus concave down on $(-\infty, 0)$ and concave up on (0∞). Thus we obtain

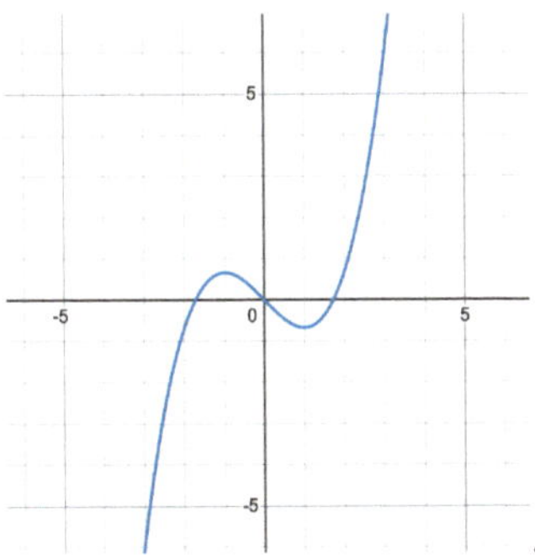

Example 4.16 Sketch the graph of

$$f(x) = \ln x.$$

Solution 4.16 We have

$$f'(x) = \frac{1}{x}$$

thus for $x < 0$, f is decreasing but this makes no sense since we are not defined for negative values and for $x > 0$, f is increasing thus increasing the whole time!

$$f^{''} = -\frac{1}{x^2}$$

thus no matter what, $f^{''} < 0$ thus the graph is concave down the whole time!

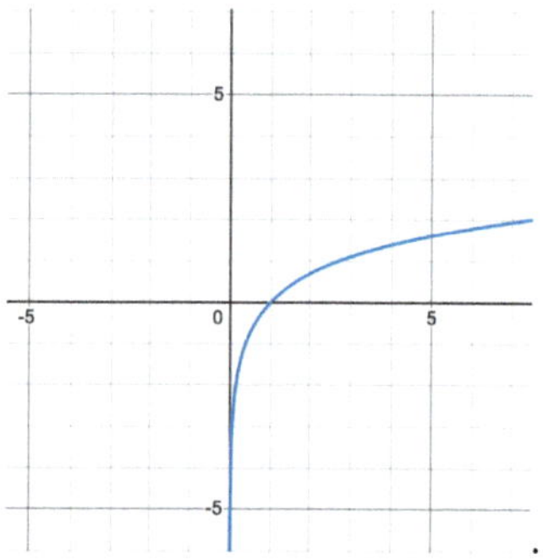

.

>§ 4.5: Exercises

1. Sketch
$$f(x) = \frac{2}{x-1}.$$

2. Sketch
$$f(x) = \frac{2x^2}{3x^2+2}.$$

3. Sketch
$$f(x) = x^3 - x.$$

4. Sketch
$$f(x) = \sqrt{x}.$$

>§ 4.6: Optimization

Often times we would like to know the maximal possible quantity obtainable under certain conditions. For instant, in a course in Algebra, one learns how to use formulas for finding the vertex of a parabola to find the maximum point of say a ball thrown. For instance, if

$$h(t) = -t^2 + 4t$$

is the formula for the trajectory of a football thrown, then the maximum occurs at the point

$$(-\frac{b}{2a}, f(-\frac{b}{2a})).$$

But we could have also found this out using calculus, namely,

$$h'(t) = -2t + 4.$$

So the function has critical number at $t = 2$ which is the number of seconds which pass for the ball to attain its maximum height. And to find this max,

$$h(2) = -(2)^2 + 4(2) = 4.$$

Lets say you have a certain amount of material to enclose a fence with one side which needs not fencing. If x, y are the sides with total fencing length say M, then we know

$$2x + y = M$$

since this would be the perimeter and we wish to maximize

$$A = xy$$

so we solve for x or y in the perimeter formula and plug this into A and differentiate A find its critical points and this will be the length of one side for optimal area.

Example 4.17 If you're given 160 feet of fencing to enclose a region in which one side needs no fencing, what dimensions maximize the area and what is this maximum area?

Solution 4.17 Let x, y be the dimensions, then we know, by formula you for perimeter, that

$$2x + y = 160.$$

And we wish to maximize

$$A = xy.$$

So from the perimeter formula,

$$y = -2x + 160$$

thus

$$A = x(-2x + 160) = -2x^2 + 160x.$$

And this has critical numbers where

$$A' = -4x + 160$$

is zero. This means $x = 40$ thus $y = -2(40) + 160 = 80$. So the dimensions are 40, 80 and so the maximum area is

$$A = (40)(80) = \boxed{3200}.$$

Example 4.18 Two integers have a difference of 8. What two numbers would you need for their product to be a minimum?

Solution 4.18 Let x, y be the two integers. We know

$$x - y = 8.$$

And their product is

$$P = xy.$$

This is what we wish to minimize. From the formula of their difference,

$$y = x - 8$$

thus

$$P = xy = x(x - 8) = x^2 - 8x.$$

And the critical numbers are those x such that

$$P' = 2x - 8 = 0.$$

This implies $x = 4$ thus $y = 4 - 8 = -4$ and this minimum product is

$$P = (4)(-4) = \boxed{-16}.$$

>§ 4.6: Exercises

1. Find two integers whose sum is 300 such that their product is a maximum.

2. You're given 45 cubic meters of box material to make a box with no top with a square base. Find the dimensions of h and w which will maximize the volume. What is this maximum volume?

3. A fence is used along all sides of a rectangular region except for one side. If the y side has one and the x has two and the total length of fencing is 160 yards, find the optimal dimensions for x, y which will maximize the area enclosed!

>§ 5. Integration

>§ 5.1: The Anti-derivative & The Indefinite integral

Remember back when we learned how to add, they then taught us how to subtract, which we disliked. Then they taught multiplication which was not too bad but it was followed shortly thereafter by division. In mathematics we have operations and then we have operations which undo these first operations. For our class now, we have learned the forward operation of *differentiation.* Now it is time to go backwards with a process known as *anti-derivation* or *integration.*

Definition 5.1: A function F is an *anti-derivative* of f on an interval I if

$$F'(x) = f(x)$$

for all $x \in I$. For instance the integral of x^2 is $\frac{1}{3}x^3$ since

$$\frac{d}{dx}\left[\frac{1}{3}x^3\right] = x^2.$$

But there is more to integrating since we may have had a constant term which became zero after differentiating, we explain this in the next theorem:

Theorem § 5.1: Representation of Anti-Derivatives If F is an anti-derivative of f on an interval I, then G is an anti-derivative of f on I if and only if

$$G(x) = F(x) + C$$

wh"ere C is a constant, known 'the *constant of integration.*

As for with derivatives how they correspond to a rate of change, an integral represents a given area. Typically the area being found is the area between the curve and the $x-$axis. The problem with taking an integral is we did not know if there was a number there and became zero once we differentiated. This is what is implies by theorem 5.2. For instance,

$$x + 1$$

and

$$x$$

both have derivative 1. This is why every time we integrate we tag on a $+C$ to account for the constant which may have been there prior to our differentiating. We write

$$F(x) = \int f(x)dx,$$

which is read *the integral of f with respect to x.* The method for constructing it was at first simple, below is the graph of

$$f(x) = x \sin x; \qquad \text{for } x \in [0, \pi],$$

(which we will not learn how to integrate until calculus 2):

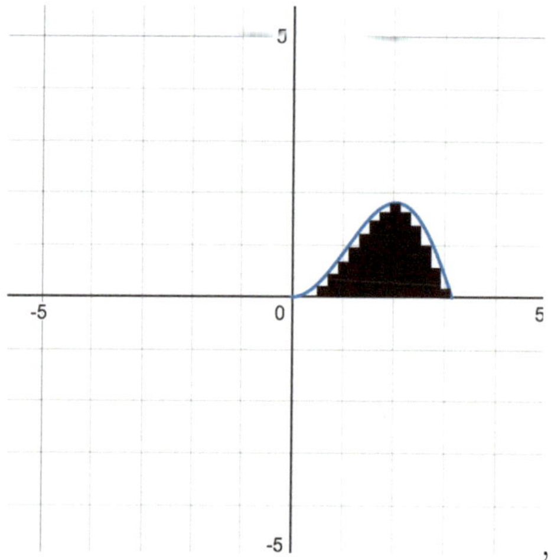

,

The idea was to draw rectangles of width dx which means *change in* x, which fit just up until under the curve, then we draw rectangles which barely pass the curve and the difference in the areas of the outside rectangles from inside rectangles is the marginal error in this inaccurate method for integrating and we call this method the method of Riemann sums. As you can see in our figure above, the rectangles leave a bit of room between their boundary and the actual curve and then the over all photo would be

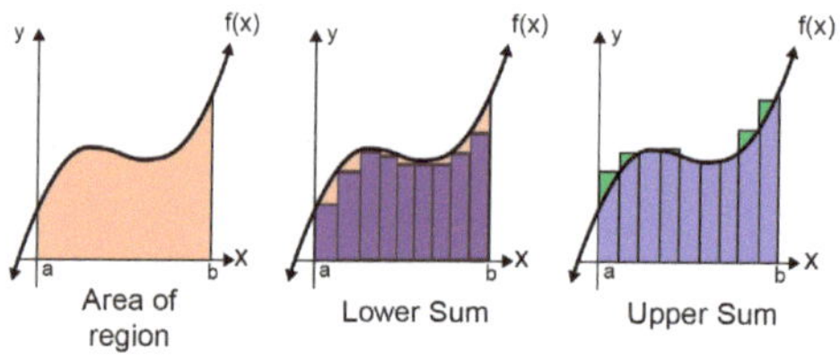

,

So the region on the far left is the solution to $\int f(x)dx$. The only rule integrals obey is the power rule:

Theorem § 5.2: Power rule for integration If you're given

$$\int ax^n dx,$$

except when $n = -1$, it becomes

$$\frac{a}{n+1}x^{n+1} + C.$$

Example 5.1 Integrate

$$\int 3x^2 + 2x + x + 1dx.$$

Solution 5.1 We get

$$\begin{aligned}\int 3x^2+2x+1dx &= \frac{3}{2+1}x^{2+1}+\frac{2}{1+1}x^{1+1}+\frac{1}{0+1}x^{0+1}+C\\ &= \boxed{x^3+x^2+x+C}.\end{aligned}$$

If we have fractions or negative exponents we rewrite them. As in the following example:

Example 5.2 Integrate

$$\int \frac{2}{\sqrt{x}}dx.$$

Solution 5.2 We have

$$\begin{aligned}\int \frac{2}{\sqrt{x}}dx &= \int 2x^{-\frac{1}{2}}dx\\ &= \frac{2}{-\frac{1}{2}+\frac{1}{2}}x^{-\frac{1}{2}+\frac{1}{2}}+C\\ &= 4x^{\frac{1}{2}}+C\\ &= \boxed{4\sqrt{x}+C}.\end{aligned}$$

In this last example, if we had a coefficient of $\frac{1}{2}$ instead of just 2 we would end up with

$$\sqrt{x}+C.$$

Theorem § 5.3: Rules for integration We have that if $k \in \mathbb{R}$, and f, g are integrable functions, then

1. $\int kf(x)dx = k\int f(x)dx$.
2. $\int[f(x)\pm g(x)]dx = \int f(x)dx \pm \int g(x)dx$.
3. $\int kdx = kx + C$.
4. $\int 0dx = C$.
5. $\int e^x dx = e^x + C$.
6. $\int \frac{1}{x}dx = \ln|x| + C$. This is the $n = -1$ case of the power rule.
7. $\int \sin xdx = -\cos x + C$.
8. $\int \cos xdx = \sin x + C$.
9. $\int \sec xdx = \ln|\sec x + \tan x| + C$. (You will not be required to show this formula as the trick in integrating is highly counter intuitive)
10. $\int \tan xdx = -\ln|\cos x| + C$. (We will learn how to actually integrate this later in this chapter.)
11. $\int \cot xdx = \ln|\sin x| + C$. (We will learn how to actually integrate this later in this chapter.)

The reason for the absolute value in (6) is that natural log cannot take on negative or zero values so we stick absolute values to ensure we have no domain issues.

Side-note: A few functions which we will not yet learn how to integrate will be:

$$\ln x, x\sin x, e^x \sin x, \frac{1}{x^2-1}, \ldots$$

Theorem § 5.4: Rules for integrating inverse trig function We have the following, for $a, b, k \in \mathbb{R}$:

1. $\int \frac{k}{\sqrt{a^2-(bx)^2}}dx = \frac{k}{b}\sin^{-1}(\frac{bx}{a}) + C.$

2. $\int \frac{k}{\sqrt{a^2-(bx)^2}}dx = -\frac{k}{b}\cos^{-1}(\frac{bx}{a}) + C.$

3. $\int \frac{k}{a^2+(bx)^2}dx = \frac{k}{b}\tan^{-1}(\frac{bx}{a}) + C.$

A few tricks for natural log, exponential, and trig functions (th following are unofficial formulas who were in some sense derived using a sort of backwards chain rule technique):

1. If $I = \int \frac{1}{ax+b}dx$, then
$$I = \frac{1}{a}\ln|ax+b| + C.$$

2. If $I = \int e^{ax+b}dx$, then
$$I = \frac{1}{a}e^{ax+b} + C.$$

3. If $I = \int \sin(ax+b)dx$, then
$$I = -\frac{1}{a}\cos(ax+b) + C.$$

4. If $I = \int \cos(ax+b)dx$, then
$$I = \frac{1}{a}\sin(ax+b) + C$$

Example 5.3(a) Integrate

$$\int \frac{2}{\sqrt{9-4x^2}}dx.$$

Solution 5.3(a) By Theorem 5.4.1, $k = 2, a = 3, b = 2$, so

$$\begin{aligned}\int \frac{2}{\sqrt{9-4x^2}}dx &= \frac{2}{2}\sin^{-1}(\frac{2x}{3}) + C\\ &= \boxed{\sin^{-1}(\frac{2}{3}x) + C}.\end{aligned}$$

Example 5.3(b) Integrate

$$\int \frac{5}{36+9x^2}dx.$$

Solution 5.3(b) We have that, $k = 5, a = 6, b = 3$ thus

$$\int \frac{5}{36+9x^2}dx = \boxed{\frac{5}{3}\tan^{-1}(\frac{x}{2}) + C}.$$

Example 5.4 Integrate

$$\int \cos x - 2\sqrt{x}dx - \frac{1}{x}dx.$$

Solution 5.4 We have

$$\begin{aligned}\int \cos x - 2\sqrt{x} - \frac{1}{x}dx &= \int \cos x dx - 2\int \sqrt{x}dx - \int \frac{1}{x}dx \\ &= \sin x - 2\frac{1}{\frac{1}{2}+1}x^{\frac{1}{2}+1} - \ln|x| + C \\ &= \boxed{\sin x - \frac{4}{3}x^{\frac{3}{2}} - \ln|x| + C}.\end{aligned}$$

Example 5.5 Integrate

$$\int \frac{1}{x^2}dx.$$

Solution 5.5 We have

$$\begin{aligned}\int \frac{1}{x^2}dx &= \int x^{-2}dx \\ &= \frac{1}{-2+1}x^{-2+1} + C \\ &= \boxed{-\frac{1}{x} + C}.\end{aligned}$$

Example 5.6 Integrate

$$\int e^x - \sin x + x + 4dx.$$

Solution 5.6 We have

$$\int e^x - \sin x + x + 4dx = \boxed{e^x + \cos x + \frac{1}{2}x^2 + 4x + C}.$$

>§ 5.1 Exercises

1. Integrate
$$\int 4x^2 - 2dx.$$
2. Integrate
$$\int x^3 - 2x.$$
3. Integrate
$$\int 4\cos x dx.$$
4. Integrate
$$\int e^x - \frac{2}{x}dx.$$
5. Integrate
$$\int \frac{1}{x^4}dx.$$
6. Integrate
$$\int \sqrt[3]{x}dx.$$
7. Integrate
$$\int x^{\frac{5}{2}}dx.$$
8. Integrate
$$\int \frac{3}{\sqrt{36-4x^2}}dx.$$
9. Integrate
$$\int \frac{2}{x}dx.$$

>§ 5.2: Definite integrals & The Fundamental theorem of Calculus

Definition 5.2 Recall from 5.1, an *integral* is the general area under a given curve called the *definite* integral and is given by

$$\int f(x)dx \qquad (*).$$

So far we always add on a $+C$ after integrating to account for this constant which may have been there. However,something like $\int_a^b f(x)dx$, is a *definite integral* whereas the one in $(*)$ we merely refer to as an *indefinite integral.* Regions above $x-$axis have positive value and below it have a negative value. In this case we must use our "evaluating" skills to determine the precise numerical answer.

Theorem § 5.5: The Fundamental Theorem of Calculus (I) Let f be continuous and real valued on $[a, b]$ and put

$$F(x) = \int_a^x f(t)dt.$$

Then F is uniformly continuous on $[a, b]$ and differentiable on (a, b) and

$$F'(x) = f(x).$$

(II) Let f be continuous on $[a, b]$ with anti-derivative $F(x)$, then

$$\int_a^b f(x)dx = F(b) - F(a).$$

Theorem § 5.6: Properties of definite integrals If f, g are integrable functions and $a, b, c, k \in \mathbb{R}$, then

1. $\int_a^c f(x)dx + \int_c^b f(x)dx = \int_a^b f(x)dx.$
2. $\int_a^b f(x)dx = -\int_b^a f(x)dx.$
3. $\int_a^b [kf(x) \pm g(x)]dx = k\int_a^b f(x)dx \pm \int_a^b g(x)dx.$
4. If f is even, then
$$\int_{-a}^a f(x)dx = 2\int_0^a f(x)dx.$$
5. If f is odd, then
$$\int_{-a}^a f(x)dx = 0.$$

As a consequence of Theorem 5.6.5,

$$\int_{-9}^9 x^5 + x^3 + xdx = 0.$$

Example 5.7 Determine

$$\int_1^4 \frac{1}{2\sqrt{x}} dx.$$

Solution 5.7 We have, by the fundamental theorem, that

$$\begin{aligned}\int_1^4 \frac{1}{2\sqrt{x}} dx &= \int_1^4 2x^{-\frac{1}{2}} dx \\ &= \frac{1}{2}\frac{1}{\frac{1}{2}}x^{-\frac{1}{2}+1}\Bigg|_1^4 \\ &= \sqrt{x}\Bigg|_1^4 \\ &= \sqrt{4} - \sqrt{1} \\ &= 2 - 1 \\ &= \boxed{1}.\end{aligned}$$

Example 5.8 Determine

$$\int_0^1 x^2 - 1 dx.$$

Solution 5.8 We have

$$\begin{aligned}\int_0^1 x^2 - 1 dx &= \frac{1}{3}x^3 - x\Bigg|_0^1 \\ &= \frac{1}{3} - 1 \\ &= \boxed{-\frac{2}{3}}.\end{aligned}$$

Example 5.9 If

$$\int_0^3 f(x)dx = 4$$

and

$$\int_3^5 f(x)dx = 5,$$

then

$$\int_0^5 f(x)dx$$

is how much?

Solution 5.9 Here we use the fact that

$$\int_a^b f(x)dx = \int_a^c f(x)dx + \int_c^b f(x)dx.$$

So

$$\begin{aligned}\int_0^5 f(x)dx &= \int_0^3 f(x)dx + \int_3^5 f(x)dx \\ &= 4+5 \\ &= \boxed{9}.\end{aligned}$$

Example 5.10 Evaluate

$$\int_{-4}^{4} 3x^2 dx.$$

Solution 5.10 Since the function is even,

$$\begin{aligned}\int_{-4}^{4} 3x^2 dx &= 2\int_0^4 3x^2 dx \\ &= 2x^3\Big|_0^4 \\ &= \boxed{128}.\end{aligned}$$

Example 5.11 Evaluate

$$\int_0^{\frac{\pi}{4}} 2\cos x dx.$$

Solution 5.11 We have

$$\begin{aligned}\int_0^{\frac{\pi}{4}} 2\cos x dx &= 2\sin x\Big|_0^{\frac{\pi}{4}} \\ &= 2\sin\frac{\pi}{4} - 0 \\ &= \boxed{\sqrt{2}}.\end{aligned}$$

When integrating absolute value functions, we set the function equal to zero and the place where the function is zero is where we split the domain of integration. I.e.,

$$\int_a^b |x - x_0| dx = \int_a^{x_0} -x + x_0 dx + \int_{x_0}^b x - x_0 dx.$$

More formally:

Theorem § 5.7: Integrals of the absolute value Suppose

$$I(x) = \int_a^b f(x)dx,$$

is the integral of an absolute value function $f(x)$. Find the root or zero of $f(x)$. That is, put

$$f(x) = 0$$

and solve. Call this number y, then

$$I(x) = \int_a^y -f(x)dx + \int_y^b f(x)dx$$

Example 5.12 Evaluate

$$\int_{-1}^{2} |x-1|dx.$$

Solution 5.12 Here, we know if

$$|x-1| = 0,$$

then $x = 1$. Thus we have

$$\begin{aligned}
\int_{-1}^{2} |x-1|dx &= \int_{-1}^{1} -x+1dx + \int_{1}^{2} x-1dx \\
&= -\frac{x^2}{2} + x\Big|_{-1}^{1} + \frac{x^2}{2} - x\Big|_{1}^{2} \\
&= (-\frac{1}{2}+1) - (-\frac{1}{2}-1) + (2-2) - (\frac{1}{2}-1) \\
&= \boxed{\frac{5}{2}}.
\end{aligned}$$

Side-Note: Anytime you are finding the area which is underneath the $x-$axis, the numerical value will be negative. As in our next example!

Example 5.13 Evaluate

$$\int_0^2 x^2 - 2xdx.$$

Solution 5.13 Here, we have

$$\begin{aligned}
\int_0^2 x^2 - 2xdx &= \frac{1}{3}x^3 - x^2\Big|_0^2 \\
&= \frac{8}{3} - 4 \\
&= \boxed{-\frac{4}{3}}.
\end{aligned}$$

>§ 5.2 Exercises

1. Evaluate
$$\int_{-1}^{2} 3x^2 dx.$$

2. Evaluate
$$\int_{0}^{4} 2x + 1dx.$$

3. Evaluate
$$\int_{1}^{e} \frac{1}{x} dx.$$

4. If
$$\int_{1}^{4} f(x)dx = 3,$$
and
$$\int_{4}^{6} f(x)dx = 4,$$
find
$$\int_{1}^{6} f(x)dx.$$

5. Evaluate
$$\int_{-1}^{3} |x-2|dx.$$

6. Evaluate
$$\int_{-2}^{1} |x+1|dx.$$

7. Evaluate
$$\int_{0}^{\frac{\pi}{4}} 2\sin x dx.$$

8. Evaluate
$$\int_{0}^{1} e^x dx.$$

9. Evaluate
$$\int_{1}^{9} \frac{1}{2\sqrt{x}} dx.$$

10. Evaluate
$$\int_{-5}^{5} x^3 - xdx.$$

>§ 5.3: Integration via u-substitution

From chapter 4, if we have a composition of functions, say

$$h(x) = f(g(x)),$$

then by the *chain rule for derivatives*, one has

$$\frac{d}{dx}[f(g(x))] = f'(g(x))g'(x).$$

And by our definition of anti-derivative in section 5.1,

$$\int f'(g(x))g'(x)dx = f(g(x)) + C.$$

This leads us to the following theorem which allows us to integrate composition functions where one of the two functions in the product is the derivative of the inside of the other function:

Theorem § 5.7 Let g be a function with range some interval, say I and suppose f is continuous on I. If g is differentiable on its domain and F is the anti-derivative of f on I, then

$$\int f(g(x))g'(x)dx = F(g(x)) + C,$$

where if we set

$$u = g(x),$$

then

$$du = g'(x)dx,$$

forcing

$$dx = \frac{1}{g'(x)}du,$$

and so

$$\int f(u)du = F(u) + C.$$

A proof of this can be done by the interested reader. ***Hint:*** *Use the chain rule.*

Example 5.14(a) Integrate

$$\int \frac{2x}{x^2+1}dx,$$

by $u-$substitution.

Solution 5.14(a) Here, our $f(g(x)) = \frac{1}{x^2+1}$ and the $g'(x) = 2x$ so we let

$$u = x^2 + 1, du = 2xdx \Rightarrow dx = \frac{1}{2x}du.$$

Thus

$$\int \frac{2x}{x^2+1}dx = \int \frac{\cancel{2x}}{u}\frac{1}{\cancel{2x}}du$$
$$= \int \frac{1}{u}du$$
$$= \ln|u| + C.$$

But wait, we had that $u = x^2 + 1$ thus

$$\int \frac{2x}{x^2+1}dx = \boxed{\ln(x^2+1) + C}.$$

And here, we can drop the absolute value bars since $x^2 + 1$ is never negative.

Example 5.14(b) Integrate

$$\int (x^3 + 2x)^4(3x^2 + 2)dx,$$

by $u-$substitution.

Solution 5.14(b) Here, $f(g(x)) = (x^3 + 2x)^4, g'(x) = 3x^2 + 2$ thus we can set

$$u = x^3 + 2x, du = 3x^2 + 2dx \Rightarrow dx = \frac{1}{3x^2+2}du.$$

Thus

$$\int (x^3 + 2x)^4(3x^2 + 2)dx = \int (u)^4\cancel{(3x^2+2)}\frac{1}{\cancel{3x^2+2}}dx$$
$$= \int u^4 du$$
$$= \frac{1}{5}u^5 + C.$$

But we let $u = x^3 + 2x$ thus

$$\int (x^3 + 2x)^4(3x^2 + 2)dx = \boxed{\frac{1}{5}(x^3 + 2x)^5 + C}.$$

Example 5.14(c) Integrate

$$\int \frac{x}{(x^2+2)^2}dx,$$

by $u-$substitution.

Solution 5.14(c) Here, we have

$$f(g(x)) = \frac{1}{(x^2+2)^2}, g'(x) = 2x.$$

Thus we can set

$$u = x^2 + 2, du = 2xdx \Rightarrow dx = \frac{1}{2x}du.$$

And so we have

$$\begin{aligned}\int \frac{x}{(x^2+2)^2}dx &= \int \frac{\cancel{x}}{u^2}\frac{1}{2\cancel{x}}du \\ &= \frac{1}{2}(-\frac{1}{u}) + C \\ &= \boxed{-\frac{1}{2(x^2+2)} + C}.\end{aligned}$$

Example 5.15 Integrate

$$\int x\sqrt{x-1}dx.$$

Solution 5.15 Here we set

$$u = x - 1, du = dx.$$

But if $u = x - 1$, then $x = u + 1$ and so

$$\begin{aligned}\int x\sqrt{x-1}dx &= \int (u+1)\sqrt{u}du \\ &= \int (u+1)u^{\frac{1}{2}}du \\ &= \int u^{\frac{3}{2}} + u^{\frac{1}{2}}du \\ &= \frac{1}{\frac{3}{2}+1}u^{\frac{3}{2}+1} + \frac{1}{\frac{1}{2}+1}u^{\frac{1}{2}+1} + C \\ &= \frac{2}{5}u^{\frac{5}{2}} + \frac{2}{3}u^{\frac{3}{2}} + C \\ &= \boxed{\frac{2}{5}(x-1)^{\frac{5}{2}} + \frac{2}{3}(x-1)^{\frac{3}{2}} + C}.\end{aligned}$$

Example 5.16 Integrate

$$\int e^{2x+1}dx.$$

Solution 5.16 We set

$$u = 2x + 1, du = 2dx$$

thus

$$dx = \frac{1}{2}du$$

and so

$$\begin{aligned}\int e^{2x+1}dx &= \int \frac{1}{2}e^{u}du \\ &= \frac{1}{2}\int e^{u}du \\ &= \frac{1}{2}e^{u} + C \\ &= \boxed{\frac{1}{2}e^{2x+1} + C}.\end{aligned}$$

>§ 5.3 Exercises

1. Integrate
$$\int \frac{2x}{2x^2+4}dx.$$

2. Integrate
$$\int (x^2+6x)^8(2x+6)dx.$$

3. Integrate
$$\int e^{3x-1}dx.$$

4. Integrate
$$\int x\sqrt{2x+1}dx.$$

5. Integrate
$$\int \frac{2x}{(x^2+9)^4}dx.$$

6. Integrate
$$\int \frac{1}{2x-1}dx.$$

7. Integrate
$$\int \frac{x}{x^2+1}dx.$$

>§ 5.4: L'Hopitals Rule

Often times a limit at first will give you $\frac{0}{0}, \frac{\pm\infty}{\pm\infty}$. In these cases, if you're told to find a given limit of some rational, and either

$$\lim_{x\to x_0}\frac{f(x)}{g(x)}=\frac{0}{0},$$

or if

$$\lim_{x\to x_0}\frac{f(x)}{g(x)}=\frac{\pm\infty}{\pm\infty},$$

then there is a little trick associated:

Theorem § 5.8: L'Hopitals Rule In the event a limit is $\frac{0}{0}$ or $\frac{\pm\infty}{\pm\infty}$, then you have that

$$\lim_{x\to x_0}\frac{f(x)}{g(x)}=\lim_{x\to x_0}\frac{f^{(n)}(x)}{g^{(n)}(x)}.$$

The n in the formula is the first $n-$th derivtive which results you in a numerical value. I.e., if you differentiate twice, your $n=2$.

Side-Note: Note we are differentiating the quotient so there is no need for Quotient rule, rather we differentiate the top and bottom separately.

Example 5.15 Evaluate

$$\lim_{x\to 0}\frac{e^x-1}{x}.$$

Solution 5.15 At first by direct substitution, one gets

$$\frac{e^0-1}{0}=\frac{0}{0}.$$

So we invoke L'Hopitals rule:

$$\begin{aligned}\lim_{x\to 0}\frac{e^x-1}{x}&=\lim_{x\to 0}\frac{\frac{d}{dx}[e^x-1]}{\frac{d}{dx}[x]}\\&=\lim_{x\to 0}e^x\\&=e^0\\&=\boxed{1}.\end{aligned}$$

Now a limit such as $\lim_{x\to 0}\frac{\sin x}{x}$ isn't so bad is it? Here we would have

$$\begin{aligned}\lim_{x\emptyset 0}\frac{\sin x}{x}\lim_{x\to 0}&\frac{\frac{d}{dx}[\sin x]}{\frac{d}{dx}[x]}\\&=\lim_{x\to 0}\cos x\\&=\cos 0\\&=\boxed{1}.\end{aligned}$$

Example 5.16 Evaluate

$$\lim_{x\to 0} \frac{e^x - x - 1}{x^2}.$$

Solution 5.16 Here, by direct substitution we obtain

$$\frac{e^0 - 0 - 1}{(0)^2} = \frac{0}{0}.$$

Thus we invoke L'Hopital:

$$\begin{aligned}
\lim_{x\to 0} \frac{e^x - x - 1}{x^2} &= \lim_{x\to 0} \frac{\frac{d}{dx}[e^x - x - 1]}{\frac{d}{dx}[x^2]} \\
&= \lim_{x\to 0} \frac{e^x - 1}{2x} \\
&= \frac{0}{0}.
\end{aligned}$$

Thus we must invoke L'Hopital once more:

$$\begin{aligned}
\lim_{x\to 0} \frac{e^x - 1}{2x} &= \lim_{x\to 0} \frac{\frac{d}{dx}[e^x - 1]}{\frac{d}{dx}[2x]} \\
&= \lim_{x\to 0} \frac{e^x}{2} \\
&= \frac{e^0}{2} \\
&= \boxed{\frac{1}{2}}.
\end{aligned}$$

Example 5.17 Evaluate

$$\lim_{x\to 2} \frac{x^2 - 3x + 2}{x - 2}.$$

Solution 5.17 Note that if we plug in $x = 2$ directly, we get $\frac{0}{0}$. Thus we differentiate and Apply L'Hopitals Rule:

$$\begin{aligned}
\lim_{x\to 2} \frac{x^2 - 3x + 2}{x - 2} &= \lim_{x\to 2} \frac{\frac{d}{dx}[x^2 - 3x + 2]}{\frac{d}{dx}[x - 2]} \\
&= \lim_{x\to 2} \frac{2x - 3}{1} \\
&= 2(2) - 3 \\
&= \boxed{1}.
\end{aligned}$$

Example 5.18 Evaluate

$$\lim_{x\to 0} \frac{\cos x - 1}{x}.$$

Solution 5.18 Note at an initial plug in, that we obtain $\frac{0}{0}$ thus we can invoke L'Hopitals Rule:

$$\begin{aligned}
\lim_{x \to 0} \frac{\cos x - 1}{x} &= \lim_{x \to 0} \frac{\frac{d}{dx}[\cos x - 1]}{\frac{d}{dx}[x]} \\
&= \lim_{x \to 0} -\sin x \\
&= -\sin 0 \\
&= \boxed{0}.
\end{aligned}$$

>§ 5.4: Exercises

1. Evaluate
$$\lim_{x\to 0} \frac{\sin x - x}{x}.$$

2. Evaluate
$$\lim_{x\to -1} \frac{x^2+3x+2}{x+1}.$$

3. Evaluate
$$\lim_{x\to e} \frac{\ln x - 1}{x-e}.$$

4. Evaluate
$$\lim_{x\to \frac{\pi}{2}} \frac{\cos x}{x-\frac{\pi}{2}}.$$

5. Evaluate
$$\lim_{x\to \infty} \frac{e^x}{x^2}.$$

>§ 6. Differential Equations

>§ 6.1: Separating a differential equations

Going back to differentiating implicitly, What if we had an equation with x's, y's and dx and dy's? Then one would expect to gather all x together with dx and all the y's together with the dy's and simplify integrate and get back what was differentiated to get the original equations. I.e., if you're at the step where its

$$dy = xdx,$$

or just $y' = x$.

Definition 6.1: An equation with x's, y's, and y''s is called a *differential equation.* The types we will study which you can pull apart y with dy and x with dx are called *separable.* To solve for y, we integrate:

$$\int dy = \int xdx.$$

Thus we have

$$\boxed{y = \frac{1}{2}x^2 + C}.$$

So the solution after integrating and solving for y with the $+C$ is the *general solution* to the given differential equation. A *particular solution* is one in which an initial condition is given and we can solve for the unknown constant, C. We will see these in the next section.

Example 6.1(a) Solve the differential equation

$$y' = 2x.$$

Solution 6.1(a) Note $y' = \frac{dy}{dx}$ thus

$$\begin{aligned} & y' = 2x \\ & \Rightarrow \frac{dy}{dx} = 2x \\ & \Rightarrow dy = 2xdx \\ & \Rightarrow \int dy = \int 2xdx \\ & \Rightarrow \boxed{y = x^2 + C}. \end{aligned}$$

Example 6.1(b) What if we have

$$y' = 2xy.$$

Solution 6.1(b) Here the xy are on the same side and we want for the y's and dy's on one side and same with x:

$$
\begin{aligned}
&y' = 2xy \\
&\Rightarrow \frac{dy}{dx} = 2xy && \text{as } y' = \frac{dy}{dx} \\
&\Rightarrow \frac{1}{y}dy = 2xdx && \text{divide by } y, dx \\
&\Rightarrow \int \frac{1}{y}dy = \int 2xdx && \text{how to undo the } d \text{ operator} \\
&\Rightarrow \ln|y| = x^2 + C && \text{rules for integrals} \\
&\Rightarrow e^{\ln|y|} = e^{Cx^2} && \text{undo ln with } e \\
&\Rightarrow \boxed{y = C_1 e^{x^2}} && \text{with } C_1 = e^C.
\end{aligned}
$$

Example 6.1(c) Solve the differential equation

$$e^{-x}y' = e^{-y}.$$

Solution 6.1(c) Note, $e^{-x} = \frac{1}{e^x}$ and if

$$e^{-x} = dx,$$

then

$$1 = e^x dx.$$

Thus

$$
\begin{aligned}
&e^{-x}y' = e^{-y} \\
&\Rightarrow e^{-x}\frac{dy}{dx} = e^{-y} \\
&\Rightarrow e^y \cancel{e^x} \cancel{e^{-x}} dy = e^x \cancel{e^{-y}} \cancel{e^y} dx \\
&\Rightarrow e^y dy = e^x dx \\
&\Rightarrow \int e^y dy = \int e^x dx \\
&\Rightarrow e^y = e^x + C \\
&\Rightarrow \ln(e^y) = \ln(e^x + C) \\
&\Rightarrow \boxed{y = \ln(e^x + C)}.
\end{aligned}
$$

Example 6.2 Find the general solution to the differential equation

$$y' = ye^x.$$

Solution 6.2 We have that

$$
\begin{aligned}
& y' = ye^x \\
& \Rightarrow \frac{dy}{dx} = ye^x && \text{since } y' = \frac{dy}{dx} \\
& \Rightarrow \frac{1}{y} dy = e^x dx && \text{a little algebra} \\
& \Rightarrow \int \frac{1}{y} dy = \int e^x dx && \text{undoing the } d \text{ operator} \\
& \Rightarrow \ln |y| = e^x + C && \text{integration rules for ln and } e \\
& \Rightarrow y = e^{e^x + C} && \text{undoing a natural log by raising by } e \\
& \Rightarrow \boxed{y = C_1 e^{e^x}} && \text{here, } C_1 = e^C.
\end{aligned}
$$

Example 6.3 Find the general solution to the differential equation

$$y' = \frac{x}{y}.$$

Solution 6.3 We have

$$
\begin{aligned}
& y' = \frac{x}{y} \\
& \Rightarrow \frac{dy}{dx} = \frac{x}{y} && \text{because } y' = \frac{dy}{dx} \\
& \Rightarrow ydy = xdx && \text{a little algebra} \\
& \Rightarrow \int ydy = \int xdx && \text{undoing the } d \text{ operator} \\
& \Rightarrow \frac{1}{2} y^2 = \frac{1}{2} x^2 + C && \text{integrating} \\
& \Rightarrow y^2 = x^2 + 2C && \text{a bit of algebra} \\
& \Rightarrow \boxed{y = \pm\sqrt{x^2 + C_1}} && \text{here, } C_1 = 2C
\end{aligned}
$$

>§ 6.1: Exercises

1. Find the general solution to
$$y' = e^{-y}(2x - 4).$$

2. Find the general solution to
$$y' = \frac{y^2}{x}.$$

3. Find the general solution to
$$xy' = 1.$$

4. Find the general solution to
$$y' = 2x \sec y.$$

5. Find the general solution to
$$y' = \frac{1}{xy}.$$

6. Find the general solution to
$$y' = y(2x + 1).$$

>§ 6.2: Initial condition problems

Given a differential equation,

$$Q(y)y' = P(x),$$

after one gets

$$y = f(x),$$

you will be given $y(a) = b$ which means when $x = a$, $y = b$ so then substitute in to find C!

Definition 6.4 Once you have an equation with only x, y as variables, and no C's, the solution is called the *particular solution* to the *initial valued* differential equation.

Example 6.4 Find the particular solution to

$$y' = xy,$$

if $y(0) = 2$.

Solution 6.4 We have

$$\begin{aligned} & y' = xy \\ & \Rightarrow \frac{dy}{dx} = xy \\ & \Rightarrow \frac{1}{y}dy = xdx \\ & \Rightarrow \int \frac{1}{y}dy = \int xdx \\ & \Rightarrow \ln|y| = \frac{1}{2}x^2 + C \\ & \Rightarrow y = e^C e^{\frac{x^2}{2}}. \end{aligned}$$

Thus we have

$$2 = C_1 e^0 \Rightarrow C_1 = 2$$

so

$$\boxed{y = 2e^{\frac{x^2}{2}}}$$

is our particular solution where $2 = e^C$ so $C = \ln 2$.

Example 6.5 Find the particular solution to

$$y' = \cos x,$$

if $y = (\frac{\pi}{2}) = 2$.

Solution 6.5 First, we solve the differential equation:

$$\begin{aligned}
&y' = \cos x \\
&\Rightarrow \frac{dy}{dx} = \cos x && \text{as } y' = \frac{dy}{dx} \\
&\Rightarrow dy = \cos x dx && \text{a little algebra} \\
&\Rightarrow \int dy = \int \cos x dx && \text{undoing the } d \text{ operator} \\
&\Rightarrow y = \sin x + C && \text{laws for integrating.}
\end{aligned}$$

But $y(\frac{\pi}{2}) = 2$ so

$$2 = \sin\frac{\pi}{2} + C \Rightarrow C = 1.$$

Thus the particular solution is

$$\boxed{y = \sin x + 1}.$$

Example 6.6 Find the particular solution to

$$y' = 3x^2 y,$$

if $y(-1) = 2$.

Solution 6.6 We have that

$$\begin{aligned}
&y' = 3x^2 y \\
&\Rightarrow \frac{dy}{dx} = 3x^2 y && \text{as } y' = \frac{dy}{dx} \\
&\Rightarrow \frac{1}{y} dy = 3x^2 dx && \text{a little algebra} \\
&\Rightarrow \int \frac{1}{y} dy = \int 3x^2 dx && \text{undoing the } d \text{ operator} \\
&\Rightarrow \ln y = e^{x^3} + C && \text{integrating} \\
&\Rightarrow y = C_1 e^{x^3} && \text{undoing ln.}
\end{aligned}$$

But $y(-1) = 2$ so then

$$2 = C_1 \frac{1}{e} \Rightarrow C_1 = 2e$$

and so

$$\boxed{y = 2e^{x^3+1}}.$$

>§ 6.2: Exercises

1. Find the particular solution if
$$y' = \frac{2}{x},$$
and $y(e) - 3$.

2. Find the particular solution if
$$y' = e^{-y}e^{x},$$
and $y(0) = 1$, and if $y(1) = 1$.

3. Find the particular solution if
$$-y' \cos y = 1,$$
and $y(0) = \frac{\sqrt{2}}{2}$.

>§ 6.3: Growth & Decay

We start here with a basic definition. And be aware that the roles of x, y are being swapped here with t, A instead.

Definition 6.1 A *growth* or *decay* is a solution to the differential equation

$$\frac{dA}{dt} = kA,$$

This is if it is growth, and

$$\frac{dA}{dt} = -kA,$$

for decay. So we have

$$\begin{aligned}
&\frac{dA}{dt} = kA \\
&\Rightarrow \frac{1}{A}dA = kdt \\
&\Rightarrow \int \frac{1}{A}dA = k\int tdt \\
&\Rightarrow \ln|A| = kt + C \\
&\Rightarrow A = e^{C}e^{kt} \\
&\Rightarrow A = C_1e^{kt} \qquad \text{here } C_1 = e^{C}.
\end{aligned}$$

And this is how the continuous growth or decay formula was derived. It was using calculus, since the rate at which it changes is some constant k times the input A. I.e..

$$\frac{dA}{dt} = kA,$$

or merely $A' = kA$. Here, C_1 is known as the *initial amount*, usually the amount after $t = 0$ time, k is the *rate*, and t is the amount of *time* that passes by. Lastly, A is the amount left after a given amount of time of a certain material. The formula (also known as *continuous compound)* is then

$$A = A_0e^{kt} \qquad k, \text{ rate, } A_0, \text{ initial amount, } A \text{ end amount, } t \text{ time.}$$

Example 6.7 A certain material initially has 160 mg left. After 4 hours it has just 120 mg left. Find the equation which represents this decay and how much is left after 6 hours?

Solution 6.7 It is worth noting that this exercise can be done with no calculus at all. Namely, we know

$$A_0 = 160, 120 = 160e^{4k}.$$

So

$$\frac{3}{4} = e^{4k} \Rightarrow k = \frac{\ln(.75)}{4} \approx -.07.$$

thus the equation is

$$\boxed{A = 160e^{-.07t}}.$$

And so at 6 hours, we have

$$A(6) = 160e^{(-.07)(6)} \approx \boxed{105.13\text{mg}}.$$

Example 6.8 A material has 150 mg initially and after 5 hours has 200 mg. Find the equation and determine how much it will have after 8 hours.

Solution 6.8 We know off the bat, $A_0 = 150$. Then

$$200 = 150e^{5k}.$$

Thus

$$\frac{4}{3} = e^{5k} \Rightarrow k \approx 1.06.$$

So the equation is

$$A = 150e^{1.06t}.$$

After 8 hours we will have

$$A(8) = 150e^{1.06(8)} \approx \boxed{722,617.48\text{mg}}.$$

There will be no exercises for this section as we have shown you how to come up with the growth decay formulas using calculus and done two examples which are precalc leveled.

Solutions to Exercises:

>§ 1. Prelims

1. $y = 2x + 8$.
2. $y = \frac{3}{2}x - 4$.
3. $\frac{3z^9}{x^8y^8}$.
4. $\frac{x^3}{2y^8z^{11}}$.
5. $f(0) = 0, f(2) = 6,\ f(x+h) = x^3 + 3x^2h + 3xh^2 - x - h + h^3$.
6. Domain is $(-\infty, -3] \cup [3, \infty)$.
7. Domain is $(-\infty, -2) \cup (-1, \infty)$.
8. Domain is $\mathbb{R} \setminus \{\frac{3}{2}\}$.
9. This becomes $\log_3(\frac{x^2y}{z^3})$.
10. This becomes $2\log_4 x + \log_4 y - 5\log_4 z$.
11. VA at $x = -4$ and HA at $y = 0$.
12. VA's at $x = \pm 2$, HA at $y = \frac{3}{2}$.
13. No VA or HA but slant at $x^3 \div x^2 + 1 = x - \frac{x}{x^2+1}$ thus the slant is at $y = x$.
14. $(-\infty, \frac{2}{3}] \cup [2, \infty)$. Or $\mathbb{R} \setminus (\frac{2}{3}, 2)$.

>§ 2.1 Limits & Things

1. -1.
2. 2
3. $\frac{1}{2}$.
4. e.
5. -1.
6. $\frac{1}{10}$.
7. 0.
8. DNE.
9. The appropriate δ is
$$\delta = \min\{|\ln(1-\epsilon)|, \ln(1+\epsilon)\}.$$

>§ 2.2 Continuity & One-sided limits

1. Polynomials like this are continuous everywhere.
2. Here, $x = \frac{1}{2}$ is not in our domain thus it is continuous on $\mathbb{R} \setminus \{\frac{1}{2}\}$.
3. We wish for $x^2 - 1 > 0$ but $x^2 - 1 = (x+1)(x-1)$ so $x = \pm 1$ are not in our domain. So the continuity is on
$$(-\infty, -1) \cup (1, \infty).$$
4. $\sin x$ is continuous everywhere.
5. e^x is always continuous everywhere.

>§ **2.3 Infinite valued limits**

1. DNE.
2. ∞.
3. DNE.
4. -1.
5. The first one has degree greater in numerator so $f(x) \to \infty; x \to \infty$. For the second, the degrees are equal so the limit is the 3 over the 2! For the last one, the top is lesser so the asymptote is at $y = 0$. For more information on how to determine limits *at* infinity, refer here.

>§ **2.4 Epsilon delta proofs**

1. Take $\delta = \epsilon$.
2. Take $\delta = \min\{|\epsilon^2 - 2\epsilon|, |\epsilon^2 + 2\epsilon|\}$.
3. $\delta = \min\{\frac{1}{2}, \frac{\epsilon}{4}\}$.
4. $\delta = \min\{\frac{1}{2}, \frac{\epsilon}{2}\}$.
5. $\delta = \frac{\epsilon}{2}$.
6. $\frac{\epsilon}{1+\epsilon}$

>§ **3.1 Differentiation and the tangent line**

1. $4x - 4$.
2. $x + 1$.
3. $\frac{1}{x}$.
4. $\cos x$
5. $\frac{1}{\sqrt{2x+1}}$.
6. $-\frac{1}{x^2}$.

>§ 3.2: Rules for derivatives

1. $f'(x) = x^3 + x^2 + x + 1$.
2. $f'(x) = \frac{1}{2\sqrt{x}} - \frac{4}{x^5}$.
3. $f'(x) = xe^x(x^2 + 2)$.
4. $f'(x) = \frac{9}{x}$.
5. $f'(x) = 2x\sin(x^2)$.
6. $f'(x) = 2x\sec^2(\frac{1}{2}x^2)$.
7. $f'(x) = \frac{6x^2}{\sqrt{1-4x^6}}$.
8. $f'(x) = \frac{2x-x^2}{e^x}$.
9. $f'(x) = 64x(2x^2+1)^7$.
10. $f'(x) = \frac{3x}{\sqrt{3x^2-1}}$.
11. $f'(x) = \frac{\cos(2\sqrt{2})}{\sqrt{x}}$.
12. $f'(x) = \frac{3}{2}\sqrt{x} - 1$.
13. $y = 6x - 2$.
14. $y = 2ex - e$.
15. $y = 1$.

>§ 3.3: Implicit & Related Rate

1. y' or $\frac{dy}{dx} = \frac{2x-y}{2y+x}$.
2. y' or $\frac{dy}{dx} = \frac{2x}{e^y}$.
3. y' or $\frac{dy}{dx} = -xy$.
4. We have r' or $\frac{dr}{dt} = \frac{1}{24\pi}$ inches per second.
5. a' or $\frac{da}{dt} = \frac{25}{12}$ inches per second.
6. h' or $\frac{dh}{dt} = -\frac{8}{3\pi}$.
7. $f^{(4)}(x) = 16\sin(2x)$.
8. $f''(x) = e^x(x+2)$.
9. $f'''(x) = 24x + 6$.

>§ 3.4: Derivative of inverses

1. $f^{-1}(3) = 2$ and $(f^{-1})'(2) = \frac{1}{5}$.
2. $f^{-1}(-5) = 1$ and $(f^{-1})'(1) = -3$
3. $f^{-1}(1) = e$ and $(f^{-1})'(1) = \frac{1}{e}$.
4. $f^{-1}(6) = 4$ and $(f^{-1})'(6) = \frac{4}{5}$.
5. $f^{-1}(x) = \frac{1+2x}{3x-2}$. .

>§ **4.1: Absolute Extrema**

1. We have absolute min at $(-1, 2)$ and two absolute maximums at $(0, 0)$ and at $(2, 0)$.
2. We have absolute min at $(-1, -\frac{4}{3})$ and absolute max at $(2, \frac{1}{6})$.
3. We have a min at $(\ln 2, 2 - \ln 4)$ and an absolute max at $(0, 1)$
4. Increasing on $(1, \infty)$ and decreasing on $(-\infty, 1)$.
5. Increasing on $(-\infty, 0) \cup (1, \infty)$ and decreasing on $(0, 1)$.
6. Increasing on $(\ln 2, \infty)$ and decreasing on $(-\infty, \ln 2)$.

>§ **4.2: Rolle's, Mean Value, Intermediate value**

1. Satisfies Rolle's theorem and $x_0 = 1$.
2. Does not satisfy.
3. $x_0 = 0$.
4. $x_0 = \frac{3}{2}$.
5. Just satisfy the conditions and you get the result.
6. Just satisfy the conditions and you get the result.

>§ **4.3: First & Second derivative tests**

1. $(\frac{1}{4}, \frac{7}{8})$ is a relative extrema points.
2. $(0, 0)$ relative max and $(\frac{1}{3}, -\frac{1}{54})$ relative min.
3. $(1, 0)$ relative min.
4. $(\frac{1}{e}, -\frac{1}{e})$ relative min.
5. $(-1, -\frac{1}{e})$ relative min.
6. Down on $(-\infty, \frac{1}{2})$ and up on $(\frac{1}{2}, \infty)$.
7. Up everywhere.
8. Down on $(-\infty, \frac{1}{12})$ and up on $(\frac{1}{12}, \infty)$.

9. Down on $(-\infty, -2)$ and up on $(-2, \infty)$.

>§ 4.4: Limits @ Infinity

1. 2.
2. 1.
3. ∞.
4. 0.
5. 0.
6. $-\frac{3}{2}$.

>§ 4.5: Curve Sketching

1. For these, use a graphing utility.

>§ 4.6: Optimization

1. The two integers are 150 and itself.
2. We have $l = w = \sqrt{15}$ and $h = \frac{\sqrt{15}}{2}$.
3. We get $x = 40, y = 80$.

>§ 5.1: Indefinite integrals & Anti-derivatives

1. $\frac{4}{3}x^3 - 2x + C$.
2. $\frac{1}{4}x^4 - x^2 + C$.
3. $4\sin x + C$.
4. $e^x - 2\ln|x| + C$.
5. $-\frac{1}{x^3} + C$.
6. $\frac{3}{4}x^{\frac{4}{3}} + C$.
7. $\frac{2}{7}x^{\frac{7}{2}} + C$.
8. $\frac{3}{2}\sin^{-1}(\frac{x}{3}) + C$.
9. $2\ln|x| + C$.

>§ 5.2: Definite integrals & The Fundamental theorem of Calculus

1. 9.
2. 20.
3. 1.

4. 7.
5. 5.
6. $\frac{5}{2}$.
7. $-\sqrt{2}+2$.
8. $e-1$.
9. 2.
10. 0.

>§ **5.3: u-Substitution**

1. $\frac{1}{2}\ln(2x^2+4)+C$.
2. $\frac{1}{9}(x^2+6x)^9+C$.
3. $\frac{1}{3}e^{3x-1}+C$.
4. $\frac{1}{10}(2x+1)^{\frac{5}{2}}-\frac{1}{6}(2x+1)^{\frac{3}{2}}+C$.
5. $-\frac{1}{3(x^2+9)^3}+C$.
6. $\frac{1}{2}\ln|2x-1|+C$.
7. $\frac{1}{2}\ln x^2+1+C$.

>§ **5.4: L'Hopitals Rule**

1. 0.
2. 1.
3. $\frac{1}{e}$ or e^{-1}.
4. -1.
5. ∞.

>§ **6.1: Separable Differential equation general solutions**

1. $y=\ln(x^2-4x+C)$.
2. $y=-\frac{2}{x^2+C_1}$, where $C_1=2C$.
3. $y=\ln|x|+C$.
4. $y=\sin^{-1}(x^2+C)$.
5. $y=\pm\sqrt{2\ln|x|+C}$.
6. $y=e^{x^2+x+C}$, or $y=C_1e^{x^2+x}$, here $C_1=e^C$.

>§ 6.2: Initial Value differential equations

1. $y = 2\ln|x| + 1$.
2. $y = \ln(e^x + e - x_)$ and then $y = x$.
3. $y = \cos^{-1}(x + \frac{\pi}{4})$.

>§ Appendix 1: Select proofs

Proof of Theorem § 3.0: Differentiability implies continuity If f is differentiable at some $x = x_0$, then it is continuous here as well.

Proof. Let $f : [a, b] \to \mathbb{R}$ be differentiable at $x = x_0 \in (a, b)$. Then for any $\epsilon > 0$ one must find a $\delta > 0$ such that if

$$|x - x_0| < \delta,$$

then

$$|f(x) - f(x_0)| < \epsilon.$$

Now for any δ, with a range of $[x_0 - \delta, x_0 + \delta]$, one has that

$$x \mapsto \frac{f(x) - f(x_0)}{x - x_0}$$

is convergent and thus bounded. So there is some $M > 0$ such that

$$\frac{|f(x) - f(x_0)|}{|x - x_0|} < M.$$

This occurs on the interval $[x_0 - \delta, x_0 + \delta]$. Well then

$$|f(x) - f(x_0)| < M\delta := \epsilon$$

and as M only gets smaller when δ gets smaller and we can choose δ sufficiently small to ensure this choice works! Note, one could have also taken $h := x - x_0$ then as $h \to 0$, one has $x \to x_0$ and shown

$$\lim_{x \to x_0} \frac{f(x) - f(x_0)}{x - x_0}$$

exists and is finite. □

Proof of Theorem § 3.2: the Power rule (For simplicity, we suppose $a = 1$) If

$$f(x) = x^n.$$

then

$$f'(x) = nx^{n-1}.$$

Proof. We have $f(x) = x^n$ an we would like to show the difference quotient exists and is equal to nx^{n-1}. Thus

$$\begin{aligned}
\lim_{h \to 0} \frac{f(x+h) - f(x)}{h} &= \lim_{h \to 0} \frac{(x+h)^n - x^n}{h} \\
&= \lim_{h \to 0} \frac{\cancel{x^n} + nx^{n-1}\cancel{h} + (\text{things with an } \cancel{h}) - \cancel{x^n}}{\cancel{h}} \\
&= \lim_{h \to 0} nx^{n-1} + (\text{things with an } h) \\
&= nx^{n-1}.
\end{aligned}$$

Thus $f'(x) = nx^{x-1}$ as needed. The step to the phrase "things with an h in it" is by the binomial theorem. In that,

$$(x+y)^n = \sum_{k=0}^{n} \binom{n}{k} x^{n-k} y^k.$$

□

Proof of Theorem § 3.4: The Chain Rule Let f, g be.real valued functions such that g is differentiable at x_0 and f is differentiable at $g(x_0)$, then the composition $f(g(x_0))$ is differentiable at x_0 with derivative

$$f(g(x_0))' = f'(g(x_0))g'(x_0).$$

Proof. We use the alternate definition of the derivative:

$$\begin{aligned}(f(g(x_0))' &= \lim_{x\to x_0} \frac{f(g(x)) - f(g(x_0))}{x - x_0} \\ &= \lim_{x\to x_0} \frac{f(g(x)) - f(g(x_0))}{g(x) - g(x_0)} \cdot \frac{g(x) - g(x_0)}{x - x_0} \\ &= \lim_{x\to x_0} \frac{f(g(x)) - f(g(x_0))}{g(x) - g(x_0)} \cdot \lim_{x\to x_0} \frac{g(x) - g(x_0)}{x - x_0} \\ &= f'(g(x_0)) \cdot g'(x_0).\end{aligned}$$

□

Proof of Theorem § 4.1: The extreme value theorem A continuous function defined on closed and bounded interval has a min and max.

Definition A.1 A set X is said to be *compact* if for any arbitrary open covering, say

$$X \subset \bigcup_{a\in A} U_a,$$

there exists a finite subset $B \subset A$ such that

$$X \subset \bigcup_{a\in B} U_a.$$

Proof. Suppose $f : [a, b] \to \mathbb{R}$ is continuous. As by the Heine-Borel Theorem $[a, b]$ is compact because it is both closed and bounded, it is enough for us to show $f([a, b])$ is compact for it would be closed and bounded and thus attain a min and max. Suppose $\{V_a\}_{a\in A}$, A some arbitrary indexing set, is an open covering for $f([a, b])$. I.e.,

$$f([a, b]) \subset \bigcup_{a\in A} V_a.$$

Then since $V_a \subset f([a, b)$ is open, $f^{-1}(V) \subset [a, b]$ is open. But $[a, b]$ is compact so there is a finite subset $B \subset A$ such that

$$[a, b] \subset \bigcup_{a\in B} f^{-1}(V_a).$$

And thus

$$f([a,b]) \subset \bigcup_{a \in B} V_a$$

and we have our finite sub-cover thus $f([a,b])$ is compact and attains both a min and maximum point. □

>§ Appendix 2: Basic Point set Topology

Before we begin, the following is a practical joke phrase in the world of topology:

"Sets are not doors".

In the sense that a door is either open or it is closed never both never neither, but some sets are open, some closed, some open and closed which are referred to as *clopen*, and some neither, as you will see shortly!! In this section we give a brief overview of the basics of point set topology. We start with the most basic notion, a given set, call it X together with a set of subsets of X, say $U_a \subset X$ where we denote

$$\tau := \{U_a \subset X : U_a \text{ is open}\}.$$

Definition A.2 The pair (X, τ) is said to be a *topological space* and the members of τ we refer to as *open sets* if the following hold:

1. Both $X, \emptyset \in \tau$.
2. If $U_a \in \tau$ for all $a \in A$ some arbitrary indexing set, then $\bigcup_{a \in A} U_a \in \tau$.
3. If $U_1, U_2, ..., U_n \in \tau$, then $\bigcap_{j=1}^{n} U_j \in \tau$.

Definition A.3 We say a subset $F \subset X$ of a topological space is *closed* if its complement $X \setminus F \in \tau$ is open in X.

Definition A.4 Moreover, a set X is *open* if it is the union of open sets contained in it and is *closed* if it is the intersection of all closed sets containing it. I.e., If X is open, then

$$X = \bigcup_{U \subset X \text{ and } U \text{ is open}} U.$$

And if X is closed, then

$$X = \bigcap_{F \supset X \text{ and } F \text{ is closed}} F.$$

Some sets that are open are $(0, 1), (-\infty, 0) \cup (0, \infty)$. Some which are known to be closed are $[0, 1]$ or $(-\infty, 0] \cup [1, \infty)$. And a neither set, in the standard topology on $\mathbb{R}$ would be $[0, 1)$ and set which is both open and closed would be all of $\mathbb{R}$ or $(-\infty, \infty)$. Both open and closed is referred to as *clopen*. I.e., X is clopen if $X \in \tau$ and $X^c \in \tau$.

Theorem A.1: Properties of closed subsets of a topological space If (X, τ) is a topological space, then

1. Both $\emptyset, X$ are closed.
2. If F_a is closed for all $a \in A$ some arbitrary indexing set, then so is $\bigcap_{a \in A} F_a$.
3. If $F_1, F_2, ..., F_n$ are closed, then so is $\bigcup_{j=1}^{n} F_j$.

Proof of Theorem A.1:

Proof. Since $\emptyset$ and X are complements of one another, and they're both open, this implies they are both closed. Since the complement of the union is the intersection of the complements, i.e.,

$$\left(\bigcup_{j=1}^{n} F_j\right)^c = \bigcap_{j=1}^{n} F_j^c$$

and since the F_j^c are open, we deduce that

$$\left(\bigcup_{j=1}^{n} F_j\right)^c$$

is open thus its complement is closed. Similarly for showing (2) in Theorem A.1. is closed. □

Definition A.5(a) If (X, τ) is any topological space, and if

$$d : X \times X \to X$$

is a function such that for all $x, y, z \in X$ one has

1. $d(x, y) \geq 0$ where equality holds if and only if $x = y$. (non-negativity).
2. $d(x, y) = d(y, x)$. (symmetry).
3. $d(x, z) \leq d(x, y) + d(y, z)$. (transitivity).

In this case we say that (X, d) is a *metric space* with *metric function* d. Here, we generalized the notion of distance in 1 dimension which is the absolute value. I.e.,

$$d(x, y) = |x - y|$$

is the most standard example of a metric space with $X = \mathbb{R}$ being the reals. Just as there are different topologies, there are different metrics as well which we will not get into.

Definition A.5(b) Let (X, τ) be a topological space. We say that that $x \in Y \subset X$ is a *limit point* if for all $U \in \tau$ with $x \in U$, we have $U \setminus \{x\}$ is nonempty. In a metric space, (X, d) this means $x \in Y \subset X$ is a limit point if for any $\epsilon > 0$,

$$B_\epsilon(x) := \{y : d(x, y) < \epsilon\} \neq \emptyset.$$

Definition A.6(c) If (X, τ) is a topological space, then we denote its *interior*, denoted X° or $\text{int}(X)$ to be

$$X^\circ := \{\text{all interior points of } X\}$$

where an *interior point*, is a point in which there is an open set containing it contained in the set. And its *closure*, denoted $\bar{X}$ or $\text{cl}(X)$ to be

$$\bar{X} := X \cup X'$$

where X' is the set of limit points. The (topological) *boundary* is $\partial X = \overline{X} \setminus X^\circ$. The interior and closure are equal to the union and intersections in Definition A.4

Definition A.6 A *separation* of a topological space X is when you can write

$$X = A \cup B; \qquad \emptyset \neq A, B \in \tau, A \cap B = \emptyset.$$

As a conclusion, both A, B are closed since they are both open and complements. But this is not in the definition of separation. A topological space for which there exists no separation is said to be *connected.*

An example of a connected space is either whole of $\mathbb{R}$ or any interval subset.

Definition A.7 A topological space (X, τ) is said to be *Hausdorff*, or T_1, if for any pair of distinct points $x, y \in X$, there exists $U, V \in \tau$ disjoint such that $x \in U, y \in V$.

Definition A.8 In a metric space, open sets become very concrete and tangeable. I.e., the open sets are for lack of a better phrase of the form

$$B_\epsilon(x) := \{y \in X : d(x, y) < \epsilon\}.$$

That is theyre all points strictly within a given distance away from a fixed point.

Fact A.1 There are different topologies!

1. The *discrete topology*, where every single subset is declared to being an open set. These spaces are clearly disconnected.
2. The *indiscrete topology*, where the only declared *open* sets are the empty set and whole set itself. This is a highly connected space.
3. The *cofinite topology* where one has

$$\tau_{\text{cf}} := \{U \subset X : X \setminus U \text{ is finite}\}.$$

   In the language of Algebraic Geometry, this is also known as the *Zariski Topology.*

4. The *lower limit topology*, where the declared open sets are subsets of the real line of the form (a, b) and $[a, b)$. So the basis for this would be

$$\mathcal{B}_l := \{(a, b) : a < b\} \cup \{[a, b) : a < b\}.$$

5. Let $K = \{\frac{1}{n}\}_{n \in \mathbb{Z}^+}$. Then let $\mathbb{R}_K$ or τ_K, called the $K-$*topology* which is the topology which includes all open intervals and open intervals of the form $(a, b) \setminus K$, or more formally, the basis of the $K-$topology can be written as

$$\mathcal{B}_K := \{(a, b) : a < b\} \cup \{(a, b) \setminus K : a < b\}.$$

In a metric space, a set $U \subset X$ is said to be *open* if for any $x \in X$ there exists some $\epsilon > 0$ such that

$$B_\epsilon(x) \subset U.$$

And a subset $F \subset X$ is said to be closed if for any $\epsilon > 0$, one has

$$B_\epsilon(x) \setminus \{x\} \cap F \neq \emptyset.$$

Fact: Every metric induces a topology on a set. These spaces are referred to as "metrizable", and clearly not every topological space is metrizable meaning we cannot establish a notion of "distance" between two points. Take a 2 point set with discrete or indiscrete topologies..

Theorem A.2: Metric spaces are Hausdorff Suppose (X, d) is a metric space. Then (x, d) is Hausdorff.

Proof. Let $x, y \in X$ be distinct points. Then for any given $\epsilon > 0$, let

$$B_{\frac{\epsilon}{2}}(x), B_{\frac{\epsilon}{2}}(y)$$

be the open sets containing x, y respectively. I claim they are disjoint. Then suppose there is some $z \in B_{\frac{\epsilon}{2}}(x) \cap B_{\frac{\epsilon}{2}}(y)$. But then by triangle inequality,

$$\begin{aligned}\epsilon &= d(x, y) \\ &= d(x, z) + d(z, y) \\ &< \frac{\epsilon}{2} + \frac{\epsilon}{2} \\ &= \epsilon.\end{aligned}$$

An contradiction since ϵ cannot be lesser than itself thus the open sets are disjoint. □

Theorem A.3: Compact subsets of Hausdorff need be closed Let (X, τ) be Hausdorff, and suppose $K \subset X$ is compact. We aim to show $K \subset X$ is closed.

Proof. To show $K \subset X$ is closed, it suffices to show $X \setminus K$ is open. Let $x \in X \setminus K$. We will construct an open neighborhood of x disjoint from K. As X is Hausdorff, for each $y \in K$, there exists disjoint $U_y, V_y \in \tau$ such that $x \in U_y, y \in V_y$. But then $\{V_y\}_{y \in K}$ is an open covering for K which is compact thus there exists some finite subset $A \subset K$ such that

$$K = \bigcup_{y \in A} V_a.$$

Then take the open neighborhood of x to be

$$U = \bigcap_{y \in A} U_y.$$

Since for each y,

$$U_y \cap V_y = \emptyset,$$

we have that $U \cap K = \emptyset$ this means that $X \setminus K$ is open forcing $K \subset X$ to be closed. □

Theorem A.4: Closed subsets of compact are compact. Let (X, τ) be a compact topological space. If $K \subset X$ is closed, then K is compact.

Proof. Let $K \subset X$ be closed with X compact. Then as K is closed, $X \setminus K$ is open. Let $\{U_a\}_{a \in A}$ be an arbitrary open covering for K. This would imply that

$$\bigcup_{a \in A} U_a \cup (X \setminus K)$$

covers X which is compact. Thus there exists some finite subset $B \subset A$ such that

$$X = \bigcup_{a\in B} U_a \cup (X \setminus K).$$

Forcing

$$K \subset \bigcup_{a\in B} U_a$$

and we have a finite subcover for K and it is thus compact. □

Next we would like to discuss how much you can "pull apart" a given space. These will be given in the form of separation axioms.

Definition A.9: The following are the *Separation axioms*: A topological space (X, τ) is said to be T_1 or *Frechét* if for any two distinct points $x, y \in X$, there exists $U, V \in \tau$ with $x \in U, y \in V$ such that

$$x \notin V, y \notin U.$$

Note, here U, V may or may nt be disjoint.

A topological space is said to be T_2 is it is Hausdorff. A space is said to be T_3 or *regular* if for any $x \in X$ and $F \subset X$ closed with $x \notin F$, there exists disjoint $U, V \in \tau$ with

$$x \in U, F \subset V.$$

A topological space (X, τ) is said to be T_4 or *normal* if for any two disjoint closed subsets $E, F \subset X$, there exists disjoint $U, V \in \tau$ such that

$$E \subset U, F \subset V.$$

That is, disjoint closed sets can be separated. Given these new separation axioms, we can go further with metric space in stating they are in fact normal topological spaces.

Theorem A.5: Metric spaces are normal Let (X, d) be a metric space. Then X is a normal topological space.

Proof. Let $E, F \subset X$ be closed and disjoint. Then for each $x \in E$, there is some $\epsilon_x > 0$ such that

$$B_{\epsilon_x}(x) \cap F = \emptyset,$$

and similarly for each $y \in F$, there is some $\epsilon_y > 0$ such that

$$B_{\epsilon_y}(y) \cap E = \emptyset.$$

I claim the open disjoint sets containing E, F respectively are

$$U := \bigcup_{x\in E} B_{\frac{\epsilon_x}{2}}(x), V := \bigcup_{y\in F} B_{\frac{\epsilon_y}{2}}(y).$$

Suppose not. Then there exists some $z \in U \cap V$. I.e.,

$$d(x, z) < \frac{\epsilon_x}{2}, d(y, z) < \frac{\epsilon_y}{2}.$$

But then by triangle inequality,

$$\begin{aligned} d(x, y) &\leq d(x, z) + d(z, y) \\ &< \frac{\epsilon_x}{2} + \frac{\epsilon_y}{2} \\ &\leq \max\{\epsilon_x, \epsilon_y\}. \end{aligned}$$

So then either $x \in B_{\epsilon_y}(y)$ or $y \in B_{\epsilon_x}(x)$. But this contradicts our choice of ϵ_x, ϵ_y. Thus $U \cap V = \emptyset$ and X is normal as needed. □

We must now turn to the concept of a basis for a topology. In fact, every element of a given topology is a union of basis elements.

Definition A.10 We say $\mathcal{B}$ which is a collection of open subsets of a given space is a *basis for the topology on* X if every $U \in \tau$ is of the form

$$U := \bigcup_{B \subset X, B \in \mathcal{B}} B.$$

I.e., $\mathcal{B}$ is a basis if

1. For all $x \in X$, $x \in B$ for some $B \in \mathcal{B}$.
2. For all $B_1, B_2 \in \mathcal{B}$, and $x \in B_1 \cap B_2$, there exists some $B_3 \in \mathcal{B}$ such that

$$x \in B_3 \subset B_1 \cap B_2.$$

One can think of how a basis generated a topology much like a basis in linear algebra generates the vector space.

Example A.1 Let $X = \mathbb{R}$ find the basis.

Solution A.1 A basis is of the form

$$\boxed{\mathcal{B} := \{(a, b) : a < b \in \mathbb{R}\}}.$$

Then any $U \in \tau$ is a union of these open intervals.

What if we wish to put a topology on a given subset of the ambient topological space?

Definition A.11 Let (X, τ) be a topological space and let $Y \subset X$ be a subset. Then the *subspace topology on* Y is defined by

$$\tau_Y := \{Y \cap U : U \in \tau\}.$$

Example A.2(a) Let $Y = [-1, 1] \subset \mathbb{R} = X$ be a subspace. Determine if

$$A = \{x : |x| \in (\frac{1}{2}, 1)\} \in \tau_Y.$$

And if $A \in \tau_{\mathbb{R}}$.

Solution A.2(a) Note $A = (-1, -\frac{1}{2}) \cup (\frac{1}{2}, 1)$. If this can be written as

$$[-1, 1] \cap U : \qquad U \in \tau_{\mathbb{R}},$$

then it is open in the subspace. But note

$$[-1, 1] \cap \big(-1, -\frac{1}{2}\big) \cup \big(\frac{1}{2}, \big) = A$$

thus $\boxed{A \in \tau_Y}$ and since it is a union of open intervals in $\mathbb{R}$, $\boxed{A \in \tau_{\mathbb{R}}}$.

Example A.2(b) Again, let $Y = [-1, 1]$ and $X = \mathbb{R}$. Determine if

$$B = \{x : \frac{1}{2} < |x| \leq 1\}$$

is in τ_Y and if it is $\tau_{\mathbb{R}}$.

Solution A.2(b) Note that

$$B = \big[-1, -\frac{1}{2}\big) \cup \big(\frac{1}{2}, 1\big]$$

and thus in $\mathbb{R}$,

$$B^c = (-\infty, -1) \cup \big[-\frac{1}{2}, \frac{1}{2}\big] \cup (1, \infty).$$

So clearly, B^c is not closed in $\mathbb{R}$ thus $\boxed{B \notin \tau_{\mathbb{R}}}$. However, one can write

$$B = Y \cap \big(-2, -\frac{1}{2}\big) \cup \big(\frac{1}{2}, 2\big)$$

thus $\boxed{B \in \tau_Y}$.

Definition A.12 Let (X, τ) and (Y, σ) be two topological spaces. If there exists a mapping

$$f : X \to Y$$

such that if $V \in \sigma$, then $f^{-1}(V) \in \tau$, we say the function f is *continuous.* Note for a metric space, the definition of continuity involving open sets which we just gave is equivalent to the $\epsilon - \delta$ definition presented in the Calculus portion of our text.

Definition A.13 A continuous function which is also 1-1 and onto with continuous inverse is said to be a *homeomorphism.* If two topological spaces are homeomorphic we say they are in a topological sense the same.

Definition A.14 A *topological invariant* is a property P that is preserved under continuity. I.e., if $f : X \to Y$ is a continuous mapping of topological spaces, we say P is a topological invariant if whenever X has property P, then so does $f(X) \subset Y$. Also known as a topological property.

Theorem A.6: Equivalence of continuity. If

$$f : \mathbb{R} \to \mathbb{R}$$

is a mapping, then f is continuous in the calculus sense if and only if it is continuous in the open set sense.

Proof. First suppose the $\epsilon - \delta$ definition hold. Let $V \subset \mathbb{R}$ be open. If $f^{-1}(V)$ is empty it is open, so suppose $x \in f^{-1}(V)$, then $f(x) \in V$. But V is open, so there is an $\epsilon > 0$ such that

$$(f(x) - \epsilon, f(x) + \epsilon) \subset V.$$

By our $\epsilon - \delta$ continuity, there is some $\delta > 0$ such that

$$(x - \delta, x + \delta) \subset f^{-1}(V).$$

I.e., x is an interior point of $f^{-1}(V)$ so its open.

Suppose on the other hand that if $V \subset \mathbb{R}$ is open, then $f^{-1}(V) \subset \mathbb{R}$ is open. Fix $x \in \mathbb{R}$ and $\epsilon > 0$. Then

$$(f(x) - \epsilon, f(x) + \epsilon) \subset \mathbb{R}$$

is open. And so

$$U := f^{-1}((f(x) - \epsilon, f(x) + \epsilon)) \subset \mathbb{R}$$

is open. This is by our open set definition being assumed. Then since $x \in U$ is an interior point, there is a $\delta > 0$ such that

$$(x - \delta, x + \delta) \subset U := f^{-1}((f(x) - \epsilon, f(x) +_{\epsilon})).$$

Thus

$$f((x - \delta, x + \delta)) \subset (f(x) - \epsilon, f(x) + \epsilon)$$

which is precisely the $\epsilon - \delta$ definition. □

One can extend this proof to include two arbitrary metric spaces. Of course this makes no sense in an arbitrary topological space as there is no defined notion of distance which is essentially what you need to make sense of the calculus continuity. This is because at the end of the day, the statement

$$|x - x_0| < \delta,$$

is a statement about distances. It merely means $x - x_0$ is somewhere inside of $(-\delta, \delta)$. I.e.,

$$-\delta < x - x_0 < \delta.$$

Theorem A.7: Connectedness and compactness are topological invariants Let $(X, \tau_X), (Y, \tau_Y)$ be two topological spaces. If $f : X \to Y$ is a continuous map and X is connected and compact, then so if $f(X)$.

Proof. In showing $f(X)$ is connected, for this, we suppose towards a contradiction that $f(X)$ is disconnected, and show this must result in X being disconnected as well. Suppose that $f(X)$ is disconnected. Then there is a separation of $f(X)$. I.e., there exists, $A, B \in \tau_Y$, with $A, B \neq \emptyset$ and

$$A \cap B = \emptyset$$

and

$$A \cup B = f(X).$$

Then

$$f^{-1}(A \cup B) \subset X$$

is open by continuity of f. Moreover,

$$f^{-1}(A \cup B) = f^{-1}(A) \cup f^{-1}(B),$$

where, by continuity of f, $f^{-1}(A), f^{-1}(B) \in \tau_X$ and they are disjoint and non empty because of our constructed surjectivity. Further,

$$\begin{aligned} f^{-1}(A) \cap f^{-1}(B) &= f^{-1}(A \cap B) \\ &= f^{-1}(\emptyset) \\ &= \emptyset. \end{aligned}$$

Thus $X \subset f^{-1}(A) \cup f^{-1}(B)$ which forms a separation therefore contradicting connectedness of X. And so $f(X)$ need be connected.

Suppose next that X is compact. We aim to show $f(X)$ is compact. That is we start with an arbitrary open covering and construct a finite sub-cover. Let $\{V_a\}_{a \in A}$, A an arbitrary indexing set, be an open covering of $f(X)$. That is, for each a, $V_a \in \tau_Y$ and

$$f(X) \subset \bigcup_{a \in A} V_a.$$

Then for each a,

$$f^{-1}(V_a) \in \tau_X.$$

This is by continuity of f. Then since arbitrary unions are closed under being open,

$$\bigcup_{a \in A} f^{-1}(V_a) \in \tau_X$$

thus $\{f^{-1}(V_a)\}_{a \in A}$ is an open covering for X and so by its compactness there exists a finite sub-cover. I.e, there exists a finite subset $B \subset A$ such that

$$X \subset \bigcup_{a \in B} f^{-1}(V_a).$$

Then

$$f(X) \subset \bigcup_{a \in B} V_a := V.$$

And we have constructed a finite sub-cover, namely V, therefore $f(X)$ is compact. □

Definition A.15 A set X is said to be *bounded from above* if there exists some $M \in \mathbb{R}$ such that

$$x \leq M$$

for all $x \in X$. And *bounded from below* if there exists some $N \in \mathbb{R}$ such that

$$N \leq x,$$

for all $x \in X$. A set that is bounded both above and below is simply called *bounded.* I.e., $(-\infty, 0)$ is bounded from above and not below. The real line $(-\infty, \infty)$ is unbounded. And the set $\{\frac{1}{n}\}_{n\in\mathbb{N}}$ is bounded both from above and below, above by 1 and below by 0.

Theorem A.8: Heine-Borel If $X \subset \mathbb{R}^n$ is a subset, then X is compact if and only if X is both closed and bounded, in the standard topology on $\mathbb{R}$, that is.

The proof of this can be found anywhere at this point so we do not bother going over it.

Consequence A.1 As a consequence of the Heien-Borel Theorem

$$[a, b]$$

is compact.

Theorem A.9: Characterization of connected subsets of $\mathbb{R}$ For $I \subset \mathbb{R}$, it is connected if and only if it is an interval. I.e.,

$$(-1, 2]$$

is connected and an interval, but

$$(-1, 0) \cup (0, 1)$$

is disconnected and not an interval.

The proof of this is very detailed and requires the least upper bound property of the reals and a whole lot of technicalities, the proof can be found anywhere these days. One direction is simple, can you guess which??

Definition A.16 A *path* from some $x \in X$ to some $y \in X$ is a mapping

$$f : [0, 1] \to X,$$

that is continuous with

$$f(0) = x, f(1) = y.$$

A topological space is said to be *path-connected* if any two distinct points can be connected using a path. A path is said to be *closed* if $f(0) = f(1)$. That is, it starts and stops at the same place. Also known as a *loop.*

Theorem A.10: Path-connected implies connected If (X, τ) is path-connected, then X is connected.

Proof. Suppose X is both path-connected, and disconnected. I.e., there is a continuous mapping

$$f : [0,1] \to X = A \cup B; \quad \emptyset \neq A, B \in \tau, A \cap B = \emptyset.$$

Then since A, B are non-empty, let $a \in A$ and $b \in B$, then these are both in X and can be connected via a path. I.e.,

$$f(0) = a, f(1) = b.$$

But $A \cap B = \emptyset$ thus

$$[0,1] = f^{-1}(A) \cup f^{-1}(B)$$

is a separation of the unit interval which is connected a contradiction. □

Theorem A.11: Characterization of Hausdorff A topological space X is Hausdorff if and only if

$$D := \{(x,x) \in X \times X : x \in X\} \subset X \times X$$

is closed.

Proof. First suppose D is closed. Then $X \times X \setminus D$ is open. And if $x, y \in X$ are distinct, then $p = (x,y) \in X \times X \setminus D$. Since this is open. Thus there exists, $U, V \subset X$ open with $(x,y) \in U \times V$ so $x \in U$ and $y \in V$. Suppose lastly, that

$$U \cap V \neq \emptyset.$$

Then $z \in U \times V$ thus $(z,z) \in U \times V$ but $U \times V \cap D = \emptyset$, a contradiction.

Suppose now that X is Hausdorff. To show D is closed in $X \times X$, we show D^c is open. Let $p \in D^c$, then $p = (x,y)$ where $x \neq y \in X$. As X is Hausdorff, then there exists $U, V \subset X$ open with $x \in U, y \in V$ and $U \cap V = \emptyset$. We must show $U \times V \cap D = \emptyset$. But the intersection of the open sets in D is non empty and U, V are disjoint so D^c is open forcing D to be closed. □

Theorem A.12: Homeomorphic equivalence We have $[0,1)$ and $(0,1)$ are not homeomorphic.

Proof. Suppose they are. Then let

$$f : [0,1) \to (0,1)$$

be a homeomorphism. A brief lemma, if

$$\varphi : X \to Y$$

is a homeomorphism, then for $x \in X$, so is

$$g : X \setminus \{x\} \to Y \setminus \{\varphi(x)\}.$$

Based off this lemma, we would have that

$$\bar{f} : [0,1) \setminus \{0\} \to (0,1) \setminus \{f(0)\}$$

is still a homeomorphism. But then $[0,1) \setminus \{0\} = (0,1)$ which is still connected, but

$$(0,1) \setminus \{f(0)\} = (0, f(0)) \cup (f(0), 1)$$

forms a separation of the space. This implies the range space is thus disconnected while the open unit interval is still connected. And thus $\bar{f}$ cannot be a homeomorphism which means f was not either. Thus no such homeomorphism of $[0,1)$ and $(0,1)$ exists. □

This trick goes to show why $\mathbb{R}, \mathbb{R}^2, \mathbb{R}^3$ are not homeomorphic. In general showing that homeomorphism of $\mathbb{R}^n, \mathbb{R}^m$ happens if and only if $m = n$ is a problem of algebraic topology which we will not get too much into in this text. The path-connected stuff leads us there as a matter of fact. For instance in showing $\mathbb{R}, \mathbb{R}^2$ are not homeomorphic we merely remove the origin and this disconnects the number line but the plane is still connected. This last fact may require some proof depending on level of rigour being shown. And in showing $\mathbb{R}^2, \mathbb{R}^3$ are not homeomorphic, we again remove the origin as the plane without its origin is not simply-connected but the three dimensional reals minus a single point at the origin is still simply-connected. I.e., simply-connectedness is still a topological invariant.

Definition A.17 A *group* is a set G together with a binary operation

$$\star : G \times G \to G$$

such that for all $a, b, c \in G$

1. $(a \star b) \star c = a \star (b \star c)$.
2. There is an element $e \in G$ unique and for each $a \in G$ there is a unique $a^{-1} \in G$ so that

$$a \star a^{-1} = e = a^{-1} \star a$$

   and

$$a \star e = a = e \star a.$$

Definition A.18 If (X, τ) is a topological space and $x_0 \in X$ is a fixed point. Let f, g be two paths in X which both start and stop at x_0. A *homotopy* between two continuous function $f, g : [0,1] \to X$ is a continuous mapping

$$H : X \times [0.1] \to X$$

such that

1. $H(0, t) = x_0 = H(1, t)$ for every $t \in [0,1]$.
2. $H(x, 0) = f(x), H(x, 1) = g(x)$ for every $x \in X$.

The set of all loops based at x_0 for a given topological space forms a group known as the *first fundamental group* and is denoted

$$\pi_1(X, x_0).$$

The group structure is *concatenation of paths.* That is, if $\gamma_1, \gamma_2 \in \pi_1(X)$ are two loops, then their binary operation, we denote $\gamma_1 \star \gamma_2$ which goes from $[0,1] \to X$ is given as

$$\gamma_1 \star \gamma_2 = \begin{cases} \gamma_1(2t); t \in [0, \frac{1}{2}] \\ \\ \gamma_2(2t-1); t \in [\frac{1}{2}, 1] \end{cases}.$$

Under $\star$, $\pi_1(X)$ forms a group (exercise). The choice of x_0 is irrelevant in a path-connected space. For instance, the first fundamental group of S^1, which to refresh you, is

$$S^1 := \{(x, y) \in \mathbb{R}^2 : x^2 + y^2 = 1\}.$$

So we have as a well established fact that

$$\pi_1(S^1, (0, 1)) = \mathbb{Z}.$$

And another known one:

$$\pi_1(\mathbb{R}^n, \{\bar{0}\}) = \{e\}.$$

That is, every loop in the real Euclidean space is homotopic to the trivial one. *Higher homotopy* groups, denoted $\pi_n(X)$ also exists but are surprisingly difficult to compute so one passes to the notion of Homology and Cohomology which requires a whole other text book. One key advantage one gets from passing from $\pi_1(X)$ to $\pi_n(X)$ where $n > 1$ is that in the higher dimensional analogs the groups are always abelian though as I have previously mentioned all the Higher homotopy groups of the $n-$sphere have yet to be computed, particularly $\pi_i(\mathbb{S}^n)$ when $i > n$ for if $i < n$ the groups is trivial and if $i = n$ you get back $\mathbb{Z}$. Recall, the $n-$sphere, denoted $\mathbb{S}^n$ is sub-space of $\mathbb{R}^{n+1}$ and is defined as

$$\mathbb{S}^n := \Big\{(x_0, x_1, x_2, ..., x_n) \in \mathbb{R}^{n+1} : \sum_{j=0}^{n} x_j^2 = 1\Big\}.$$

However, one interesting fact that one can prove very easily from these objects is the infamous consequence of the *Invariance of Domain*, which states that

$$\mathbb{R}^m \cong \mathbb{R}^n$$

if and only if $m = n$. For simplicity sake, we will show That $\mathbb{R}, \mathbb{R}^2$ are not homeomorphic and a proof sketch of why $\mathbb{R}^2, \mathbb{R}^3$ are not homoemorphic.

Consequence A.2(a): $\mathbb{R} \ncong \mathbb{R}^2$.

Proof. Suppose $\mathbb{R} \cong \mathbb{R}^2$, then there exists a homeomorphism

$$f : \mathbb{R} \to \mathbb{R}^2.$$

There is a fact that if you remove a point from each space under a homeorphism, the resulting function is homeomorphic. So consider the map

$$g : \mathbb{R} \setminus \{0\} \to \mathbb{R}^2 \setminus \{f(0)\}.$$

But $\mathbb{R} \setminus \{0\} = (-\infty, 0) \cup (0, \infty)$ which forms a separation and is no longer connected. But $\mathbb{R}^2 \setminus \{f(0)\}$ is still connected which is a topological invariant. □

Definition A.19 A topological space is said to be *simply connected* if it is both path-connected and has trivial fundamental group, that is, all loops are homotopic to a point.

Consequence A.2(b): $\mathbb{R}^2 \ncong \mathbb{R}^3$.

Proof. Do the same thing, only this time when you remove a point from $\mathbb{R}^2$ it is no longer simply connected because for instance the unit circle is not contractible to a point but $\mathbb{R}^3$ minus a point is still simply connected so simple connectedness is a topological invariant. □

So dimension is a topological invariant.

We now discuss this notion of different topologies on a given set. In fact, we have a few key facts about single set multiple topology scenarios. Suppose we have one set, X and an uncountable number of topologies, τ_a, $a \in A$ an arbitrary indexing set.

Theorem A.13: Closures of topologies If X is a set and τ_a for $a \in A$ are topologies on X, then

1. $\bigcup_{a \in A} \tau_a$ need not be a topology on X.
2. $\bigcap_{a \in A} \tau_a$ is a topology on X.

Proof. To prove (2), merely follow definitions, and to see why (1) fails, consider $X = \{a, b, c\}$ and put

$$\tau_1 := \{\emptyset, X, \{a, b\}\},$$

and

$$\tau_2 := \{\emptyset, X, \{a, c\}\}.$$

Then $\{a, b\} \cap \{a, c\} = \{a\}$ but $\{a\} \notin \tau_1 \cup \tau_2$ thus the union of these two topologies is not closed under intersections. □

Theorem A.14: A compact non-Hausdorff topology Let $(\mathbb{R}, \tau_{\text{cf}})$ be the real with the cofinite topology. This space is compact and non-Hausdorff.

Proof. First suppose we wish to show it is compact. Let $\{U_a\}_{a \in A}$, A an arbitrary indexing set, be an open covering. Fix some $a \in A$, then since $U_a \in \tau_{\text{cf}}$ for every a, its complement in $\mathbb{R}$ is finite. That is,

$$\mathbb{R} \setminus U_a = \{x_1, x_2, ..., x_n\}.$$

For each j, $1 \leq j \leq n$, let U_{x_j} be the open set containing x_j. Then it is clear to see

$$U_a \cup \bigcup_{j=1}^{n} U_j$$

is a finite sub-cover of $\mathbb{R}$. As $\mathbb{R} = (\mathbb{R} \setminus U_a) \cup U_a$.

Next we wish to show it is non-Hausdorff. Let $x.y \in \mathbb{R}$, then if it was Hausdorff, there exists U, V open in $\mathbb{R}$ with $x \in U, y \in V$ and

$$\mathbb{R} \setminus U = \{p_1, p_2, ..., p_m\},$$

and

$$\mathbb{R} \setminus V = \{q_1, q_2, ..., q_n\}.$$

As $\mathbb{R}$ is infinite, there exists some z such that $z \neq p_i, q_j$ for $1 \leq i \leq m, 1 \leq j \leq n$, this forces $z \in U \cap V$ and therefore they are not disjoint thus the space is non-Hausdorff. □

> § Index:

A:

B:

C:

F:

G:

H:

I:

Thank you for reading!! Volume II: My (2) Cents on Calc (2) coming soon!! Be on the lookout!!!!

(For questions, comments, concerns, email me at: hossien.sahebjame@sjcd.edu)

>§ 11. PS- Differences between Algebra & Calculus

The key differences are:

Subject:	Algebra	Calculus
differences	average rate measure of "box" or circular region sum of finite series optimize single variable	instantaneous rate measure of varied curvature surfaces sum of infinite series optimize multi-variable

Original publish date: 11/22/2025; Pasadena, TX.

www.ingramcontent.com/pod-product-compliance
Lightning Source LLC
LaVergne TN
LVHW071241210926
848486LV00007B/321